Metapatterns

Across Space, Time, and Mind

Metapatterns

Across Space, Time, and Mind

Tyler Volk

Columbia University Press • New York

Columbia University Press
New York   Chichester, West Sussex
Copyright © 1995 Columbia University Press
All rights reserved

Library of Congress Cataloging-in-Publication Data
Volk, Tyler.
    Metapatterns across space, time, and mind / Tyler Volk.
     p. cm
    Includes bibliographical references and index.
    ISBN 0–231–06750–X
    1. Metacognition.   2. Thought and thinking.   I. Title.
    BF311.V58   1995   94–40589
    117—dc20   CIP

∞

Casebound editions of Columbia University Press books are printed on
permanent and durable acid-free paper.

Printed in the United States of America
c   10   9   8   7   6   5   4   3   2

Contents

Prologue

What Are Metapatterns?

I have borrowed the word *metapattern* from Gregory Bateson (1904–1980). The great synthesizer used it in his masterpiece, *Mind and Nature*. Oddly, the word appeared only once—though prominently—in the introduction. There Bateson, who worked in anthropology, biology, philosophy, epistemology, cybernetics, and ecology, conveyed the key to his method of thinking. He did so by way of a story.

When teaching, he would pull a crab from a bag. Then the Socratic inquiry would begin: In what ways do the two claw-equipped limbs share a common anatomy, despite differences in pincer size? Now, how do these large front limbs resemble the walking appendages? Repeat the process with a lobster. Then, how do crabs compare with lobsters? Finally, how does this generalized arthropod pattern compare with the mammalian pattern, drawn from a parallel exercise with a human and a horse? Along the way, he would urge his responders to keep in mind what he called "the discarding of magnitudes in favor of shapes, patterns, and relations."

What does one gain, what does one perceive by moving through these layers of comprehension? Patterns of patterns—metapatterns.

Bateson took this method well beyond the borders of comparative biology in his quest for the "patterns that connect" (perhaps his most remembered phrase). Consider biological evolution and human learning, for example. To Bateson they were both instances of a more inclusive pattern. Both use the metapattern of variation and selection—an explosion of possibilities followed by a culling with a strong component of effi-cacy—to forge a trail in the possibility space of new configurations. One yields forms in the biological world; the other, forms in the psyche.

I was fortunate to have studied with Bateson while he was writing *Mind and Nature*. It was autumn of 1977, and he was scholar-in-resi-dence at the Lindisfarne Association in New York City. Once every two weeks he held an all-day, free-wheeling seminar, which I could attend because of the itinerant nature and scattered schedule of my work as a freelance plumber, carpenter, and teacher. At that time I was teaching two courses at The School of Visual Arts: "Visual Science" and "Patterns in Time." The first looked at universal patterns in space (here chapters 1–6); the second, in time (chapters 7–10).

The main patterns I was playing with then—spheres, borders, arrows, breaks, and the like—survived the two decades of alternating scrutiny and inattention. The sub-metapatterns, here the sections within chapters, were still years away from making their presence known to me. And only recently have I taken Bateson's term to heart and mind as the overarching descriptor for what continues to be, for me, a family of inspiring concepts.

Initial glimpses of the metapatterns owed to a period of immersion in nature in the early 1970s, during several years of freethinking and wandering after graduating college with a degree in architecture (which I was not keen on immediately utilizing, despite a passion for the subject). Later, in the early 1980s, when I gave up my collaged career to pursue a doctorate in energy and earth sciences, the meta-patterns continued to animate and, yes, haunt me. To my delight I found that even the esoteric crannies of disciplinary science were their domain. Metapatterns helped me formulate models of the ocean's car-bon cycle and understand the structure of scientific debates. And when, in the 1990s as a tenured professor, I had time during summers and a sabbatical to concentrate on this book, the metapatterns expanded into vistas of questions I could easily spend a lifetime on.

I have offered Bateson's round-about definition of a metapattern. What of mine?

To me, a metapattern is a pattern so wide-flung that it appears throughout the spectrum of reality: in clouds, rivers, and planets; in cells, organisms, and ecosystems; in art, architecture, and politics. The third set, representing all of human creativity, is especially rich with what I perceive as metapatterns—as it should be. Images and insights that pull at my own thoughts are sure to have influenced those of others.

I use the word metapattern in the Batesonian spirit—as a pattern of patterns—and seek examples at the very broadest scale. Alas, my definition, too, is round-about. I define metapatterns by saying where they are found and how I use them. But what *are* they? And are they out there (patterns sensed) or in here (patterns imagined)?

Suppose you were asked to define a canoe. You describe a canoe's shape, its dimensions, materials, even methods of construction—as if preparing to build or at least to recognize one. In another type of answer you might describe what a canoe does, how it functions, namely, carrying a person across water. Perhaps in this case the listener might need to use the canoe.

There is yet a third way of responding. Rather than saying anything directly about the canoe, you describe the experience of being in a canoe, what can be seen while paddling around—perhaps creeks tumbling from forested gorges into a secluded lake. This third way of answering is the way I have chosen to present the metapatterns.

This book is thus a travelogue. It contains views of reality seen from the canoe of metapatterns. The various creeks I visit are the many disciplines—physics, chemistry, biology, ecology, psychology, mythology, culture. Because metapatterns have given me a canoe for exploration, they are admittedly something I have constructed, a self-developed way of thinking. But this is only part of the story, one face of a rich metaphor.

Let's say that the metapatterns are not the canoe but the lake itself. Just as the feeder streams flow into this single body of water, so too the streams from many regions of reality pour into the great reservoir of metapatterns. Perhaps the metapatterns are attractors—functional universals for forms in space, processes in time, and concepts in mind. Surely in this mind they have served as such. I invite you to let them enter into yours.

Tyler Volk

Metapatterns

Across Space, Time, and Mind

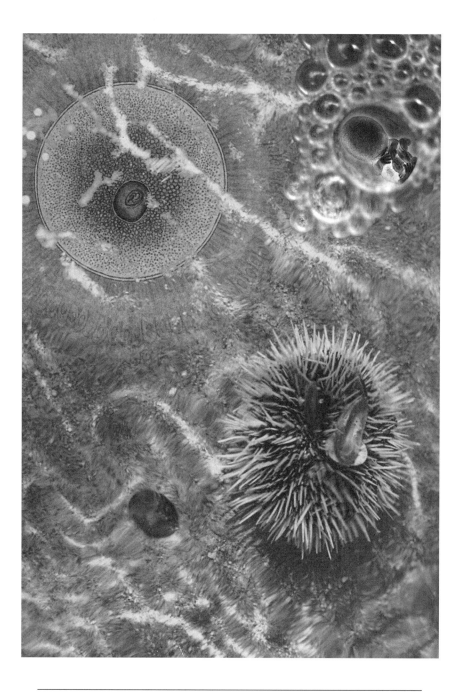

In the beginning is the sphere. This collage contains a human ovum (with its dark nucleus), a human fetus tucked inside a bubble, and a sea urchin (which is using the domelike shell of a mussel for protection against the sun at low tide.)

Spheres

We began life as simple, floating spheres. As eggs we popped from follicles in ovaries of mothers-to-be. Fertilized by sperm in fallopian tubes, dividing again and again, our spheres persisted. But when we nestled and flattened into the womb's wall, and, later, groped with arms and kicked with legs, an interplay between the sphere and its contrary began. This interplay forever follows us: by day we walk as upright sticks; at night we curl into fetal balls.

After birth and during our earliest years we were spheres of another kind. As children our minds were round. We explored in all directions, took interest in diverse everythings and anythings. But whoever can recall the early sphere of consciousness can probably recall, too, that inevitable change. Desires flattened, proclivities groped and kicked; we became adults with discriminating interests, professions, friendships.

Within, however, still resides the psychological sphere as center, as nucleus for our deepest dreams and drives. At the grand scale religion, philosophy, and science seek for the universal, the embracing sphere of experience and explanation. Closer to home we may be invigorated by a book from a field we normally avoid, by an acquaintance with whom we normally would not converse, by stories from distant lands and times that make us more well rounded, more complete and whole. Were I, in fact, held to just one theme, one message to be carried in this book, it would be a celebration of our human need and capacity to retrieve and achieve our spheres. Journeying with this theme in mind has led me to the metapatterns.

 ## Majesty of the Sphere

Our own spherical origin echoes the starting shape from which virtually all living things emerge. The sphere proclaims some original, functional, and perhaps even aesthetic truth embedded in today's burstings of life throughout the five kingdoms and even back through the eons of evolution. In the fertilized eggs of frogs, sea urchins, and worms, in the opal ovules of grape embryos, we see the sphere as the beginning of complex individuals. In the glistening green balls of freshwater algal colonies of *Volvox*, in many kinds of marine plankton—whose bulbous, protective shells are as intricate as their names—and even in cells as old as *Melasmatosphaera* (now suspended in rock, rather than in the waters of two billion years ago), we see the sphere not only as origin but as culmination of life forms.

Floating also in the ocean's planktonic layer is the tripod of a larval sea urchin, a temporary excursion in a life cycle between the sphericity of egg and adult. But the urchin's final return to roundness is rare in biology. More commonly only particular body parts retain the ancestral shape. The fertilized grape ovule, for example, grows away from sphericity as a sinuous vine; only in its fruits does it return to the primordial shape. And for the human, only heads, testes, breasts, butts, eyeballs recall the original shape.

Turning away from life and to the macro- and micro-regimes of matter, the sphere again reveals a majesty of ubiquity. Astronomers see spheres filling the cosmos, from red giants to white dwarfs, from flashy pulsars to the Hawking event horizons of black holes. The planets and their larger moons are also balls, and thus brethren to the stars. To be sure, as evidenced by Earth with its slight equatorial bulge, sphericity never reaches Euclidean perfection. Yet the tendency is strong. In our solar system only the very small bodies, astronomically speaking, such as the Manhattan-sized moons of Mars, escape the spherizing crunch of gravity's mold.

Eighty percent of the visible universe is apparently hydrogen, the first element—and the first to have its shape as an individual atom computed

Spheres at all levels. Earth, atoms, a pollen grain, a spiked radiolarian shell; underlying all (by giving us energy to build devices for seeing these spheres) are soybeans.

by quantum theory. When alone in space, like a miniature star, a hydrogen atom's cloud of charge density created by its single electron extends and fades into spherical infinity (the so-called *1s* electron: *s* for spherical). Higher-numbered clouds of larger atoms can be topologically complicated, and molecular bonding further tugs the fields of charge away from perfect sphericity. Nevertheless, as the breathtakingly detailed images of the scanning tunneling microscope now confirm, the simplified idea of atoms as space-filling balls conveys an essential truth.

The sphere thus appears to reign as dominant shape in the astronomically immense, the atomic infinitesimal, and the ancient or nascent living. No other shape is so universally abundant, so insistent as the omnipresent sphere.

The Grape-Moon Koan

We have surveyed briefly the where of the sphere, let us now stalk the why. Begin with a crystalline, full moon night. In one sweep, hand-brush a thick, black circle—an *enso* in the tradition of Zen art. Now place the moon inside the enso. Also, to playfully engage a binary, toss a grape into the ring of ink and connect it to the moon. A Zen puzzle, a visual *koan*, emerges: the grape-moon koan. Through it we are called to consider how, if at all, astronomy's moon and biology's grape relate. Are they two spheres or one?

A koan (KO-ahn) is a question or vignette that is puzzling, absurd, paradoxical, enigmatic. It has the power to evoke fundamental questions and realizations. The hoped-for burst of insight, or satori, however, must be found anew by each reflecting individual. My exploration of the grape-moon koan should therefore be taken as merely suggestive commentary. Moreover, my interpretation could be considered unfair because I indeed fabricated the koan. Well, yes, but also no—nature presented it to me, or, rather, tripped me over it and changed my life.

One early autumn day in rural New York State, I was walking through an old, overgrown field, approaching the fruit trees and garden. As I passed alongside the grape trellis, my eye caught on a ripe, luscious sphere. Much more than my eye, the grape ensnared my mind. Sky, field, vine, and all other objects of thought and attention disappeared. Some tunnel of entrancing resonance linked me to that shiny, dark purple orb—so perfectly round it seemed as much an embodied idea as a biological growth.

Later that same day, into dusk and early evening, I was wandering the hillsides—a daily soul feast in that hippie, Wordsworthian era of my life, when I was wont to ponder the architecture of nature. I was bemused by the inexplicable intensity of my identity with the grape and was lost in thought as the landscape dimmed into night. Suddenly a shadow fell before me, my own. I turned and beheld above the low hills to the east, the moon. Again, a reverie took possession of me. All disappeared, except for the moon and, significantly, the grape, which at that moment was as intense in memory as the moon in my retina. Like the celestial bodies during a solar eclipse, grape and moon coalesced into a unity, a koan.

That experience was the seed that grew into this chapter, indeed, this book. From the similarity in shape of grape and moon came the sphere as a metapattern able to bridge seemingly disparate realms—astronomy, biology, and, later for me, even psychology. Out of other seeds of experience came more metapatterns. I now see all the metapatterns as koans, as questions without single answers, as ever-fresh

The grape-moon koan.

sources of inspiration, inviting explorations of the form and function of, simply, everything.

Let us begin with the basic koan: the grape and the moon as physical spheres. Consider first a simpler pair: the moon and sun. Both are forged by the omnidirectional field of gravity. But though gravity may explain their shapes, it cannot alone account for their long-standing stabilities of size. That stability derives, rather, from the contracting effect of gravity being balanced by expanding forces. In the sun the radiation pressure of nuclear fusion blasts outward from the core. The moon's counter to gravity comes from the summed resistance of all its atoms against being squeezed any tighter. Omnidirectional forces that pattern these stable bodies of matter thus come in pairs: gravity squeezes and either a torrent of heat and light or the resistance of atoms pushes back.

A similar balance of countervailing, omnidirectional forces works the shape and size of atoms. Electric attraction squeezes electrons toward the central nucleus, while statistical fluctuations in their momentum (formalized as the uncertainty principle) fling the electrons outward in all directions.

Stars and atoms are visible maps of the invisible but controlling powers that underlie sphericity in far-flung scales of the universe. Gravity and electromagnetism on the one hand and radiation pressure, atomic rebound, and momentum fluctuations on the other are all spherically symmetric—their strengths of attraction or repulsion change equally in all directions relative to a center.

Could a grape also be shaped by a balance between contraction and expansion? A grape is, however, more than an instantaneous balance of physical forces in space. With the grape we jump into a more encompassing level of cause—one that brings in time. The grape-moon koan becomes a biology-physics koan, and even a time-space koan.

During growth a grape's volume of reproducing and swelling cells expands. So as not to become a banana or string bean, the growth of a grape is regulated to expand evenly, omnidirectionally. This regulation, taken as a whole, constrains the resulting growth to a particular shape and can be considered as a concurrent contraction that guides the expansion. Both growth and its control are highly evolved, dynamic, and complex processes in a living thing—processes rather unlike the stark simplicity of the forces of physics. Nevertheless, from a pattern viewpoint, both grape and moon are given shape by influences that are omnidirectional.

So spheres are created by omnidirectional processes . . . this is the insight? Do we not, after all, define the geometric ideal, the sphere, precisely by its omnidirectionality in space from a central point? The insight smacks of tautology. But sometimes when reason spins in a circle, it allows one to spiral ahead. Returning again and again to omnidirectional processes and sphericity has created a link, a path between the moon and sun, then to atoms, and finally to a grape.

With a well-grounded generalization one should be able to explore specifics even further along the helix of examples. Suppose some Zen-imbued trickster approaches the grape-moon koan. As prototype, consider the tale of a novice who instantly responded when the Zen master shouted this koan to the group: "Call me anything you wish, but not 'water jug'." The novice ran up and kicked over the jug, thus proving his enlightenment. So too, one might respond to the grape-moon koan by throwing a baseball into the field of the enso. Ha—answer that!

Cosmogenically speaking, baseballs only recently joined grapes and moons in the league of major spheres. An inert lump, a baseball resists

Play ball.

a contractive pulse from a bat's blow with a rebounding pressure of its fibers and ultimately its atoms. In countering a physical contraction with a physical expansion, this fruit of technology is something like the moon.

But the causes of its form are more complex and, like those of grapes compared to moons, they present yet another level of logic. Cultural desires and rules—not gravity or cell growth—prescribe sphericity to that leather-bound mass of fibers around a hard rubber core, wound in a factory by workers and machines, measured and approved by the sport's regulators. Its size is dictated by design considerations that both expand and contract, balancing advantages to the batter and the pitcher.

We require the ball to behave just so in our games. Spit on or scar a baseball, and the ump discards it from play. Fairness of bounce demands equality of extent, and so a performance requirement (itself fair and thus omnidirectional on still another level, extending to all players) determines the design. Such needs have shaped the spheres hit, hurled, and kicked across playing fields for thousands of years. Play ball!

Small Surface, Great Volume

The bat whacks, the ball flies away spinning. The catapult swishes, the boulder shoots toward an ancient city wall. The cannon booms, the solid iron globe pocks Napoleon's army. These spheres allow loading, release, and impact in any orientation, and, while in flight, they can spin with minimal air resistance. The rounding of rocks by rivers reverses the process: rather than being crafted to flow, the flow crafts them.

The first satellites—Sputnik and Vanguard—were balls for similar reasons. Their roundness minimized drag with the upper reaches of the atmosphere, and they could tumble without orientation whilst sending beeps to Earth. Despite today's more cylindrical communication satellites and streamlined tubes of space shuttles, space is still quite the place for spheres. And not just in the fiction of Darth Vader's starship; for example, consider shuttle astronaut Story Musgrave's unscheduled experiments with soda pop.

"It's still magic, even though you're used to things floating," Story told me in Houston, at NASA's Johnson Space Center. When squirt from its container, a stream of cola shapeshifts into a globule. Halting its drift through the cabin is difficult. Probe with a finger and it attaches like a leech. One must instead puff through a straw from alternating sides to effect stasis in and hence nearly perfect sphericity of the floating liquid. To see how foam would separate, Story also used a straw to set the sphere spinning. The air bubbles rose, so to speak, by descending inward, coalescing into an inner sphere at the center. Story mused about what salad dressing might do.

A cola sphere in orbit readily adheres to fingers for the same reason that it reminded Story most of an Earth-based balloon filled with water. In space, the liquid ball has a "skin" that is self-created from surface tension. On Earth, water striders skate across pools, drops build up and break from leaky faucets, and we can float razor blades and fill glasses slightly above their rims all because liquids naturally form such skins. Skins form because the fluids drive toward states of lowest total

Rolling, tumbling, shooting spheres. The giant cannonball that now rests in a Moscow park was one of the largest ever made (the cannon barrel it was meant for looms above the lion's head in this collage). A lichen-covered boulder in Australia once tumbled in a rain forest river. *Vanguard 1*, a U.S. satellite, was launched in 1958. On top of the cannonball, dung beetles roll their prize.

energy, which translates into the lowest total area of free surface. In the world of zero-g this drive works in full three-dimensional glory upon blobs of water, soda pop, or salad dressing, tightening them into the volume with the least surface. That volume is, of course, the sphere.

With this physical principle in mind, it is worthwhile to revisit the grape. A minimized surface area does have distinct biological advantages. Water loss, for example—which is a challenge for many fruits—can be reduced by emulating sphericity. This functional trait adds yet another level to the complexity of the grape compared to the moon or soda pop sphere. The grape is a product of selective evolution. Its shape thus represents a linking of metabolic functions to attributes of pure geometry.

Wherever life needs to enclose, contain, and separate by minimizing area of contact with the environment, a sphere is often the answer. Because the emphasis is on protected internal development, animal

Sphericity in space. Story Musgrave floats in the space shuttle, after experimenting with soda pop behavior. He photographed the ball of cola, about two inches long, in accelerated motion, as shown by its deviation from sphericity and by the asymmetric distribution of the bubbles.

eggs, seeds, and buds of flowers and leaves do well as spheres. Mature leaves are usually far from spherical, but when water retention becomes the overriding design objective, as in desert plants, then relatively more sphericity comes to the fore: the cacti.

At smaller scales, the area-minimizing force of surface tension in the lipid (and fluidlike) membranes of many cells snaps them into balls like soda pop spheres—the low maintenance shape. To be other than spherical, a cell must build sturdy walls or continually exert extra energy on the membranes. That famous nonspherical protist, the amoeba, becomes a sphere when it dies.

As spheres, the floating, single-celled plankton—the foraminiferans, coccolithophorids, and radiolarians—can precipitate their shells with a minimal amount of materials. The majority of cells in our immune system are orbs—for example, the T-cells, the B-cells, and the natural killer cells. Even the phagocytes bounce along as spheres until they are called upon to engulf some micro-bit of nastiness and thus need to extend their surfaces.

Purely physical examples of spheres produced by surface minimization abound on Earth. In ancient ocean sediments the giant meteor impacts of the past have left layers of glassy microtectites, once-molten droplets of rock splashed from the impact sites. Certain types of volcanoes produce spherulites, also formerly molten. More familiarly, behold the soap bubble. But at the very smallest scales (just above the atom, whose spherical electron shells are also a kind of energy minimization), the supreme example of sphericity has only recently been discovered. High-tech combustion experiments have revealed a preponderance of products with sixty atoms of pure carbon. The number sixty is no fluke (it could not be fifty-nine or sixty-one), because sixty is one of nature's magic numbers that allow packing with regular and equally distributed connections around the surface of a sphere. In chemical terms, sixty is one recipe for minimized energy. The newly discovered molecule's official name is buckminsterfullerene; its unofficial name, the buckyball.

 ## Buckyballs, Domes, and Shells

Carbon buckyballs are incredibly strong. Researchers have noted their survival after collisions at twenty thousand miles per hour—a durability which far surpasses that of other molecules. One of the codiscoverers of buckyballs, the English astrochemist Harry Kroto, had long been a fan of the creations of Buckminster Fuller—which included a geodesic sphere that Fuller designed for the U.S. pavilion at the 1967 World's Fair in Montreal (the sphere spanned 250 feet, with no internal supports). Twenty years later, with a team of chemists pondering how sixty atoms of carbon might join (the team knew the number of atoms in their newly created molecule, but not the configuration), Kroto recalled he had once constructed a model of the starry sky using hexagons and pentagons. Ultimately, the team realized that the carbon atoms dance in the geometry Fuller called the hex-pent dome. Sports fans know it as the soccer ball.

Buckminster (Bucky) Fuller not only designed sphericity; he personified it. He was geometer, architect, engineer, poet, philosopher; analyst of history, economics, and politics; orator and coiner of "spaceship Earth." Above all, Fuller will be remembered for designing the most efficient self-supporting structures ever fabricated.

In most buildings—from a sprawling shopping mall to a wood-frame shack—forces at any point will be directed into the earth by only a small percentage of the total structure. This is because columns direct the force of their loads in only one direction: straight down. By contrast, the complex curving and interwebbing of adjacent parts in domes, eggshells, virus capsids, and buckyballs spreads any local force over the entire structure. The structural integrity intimately ties the parts to the whole. Thus, for a limited mass of structure, a dome provides a maximally capacious, fuller building.

In 1957 Buckminster Fuller designed for the Union Tank Car Company in Baton Rouge, Louisiana, a repair shop using his new geodesic, spherical geometry. The shop's 384-foot diameter was the largest clear-span, self-supporting enclosure in history. This was more than double the diameter of what had held the world record for nearly two thousand years: the Roman Pantheon. Built under Hadrian's rule (for activities unknown), the Pantheon, unlike most pagan monuments, survived the Christian "ax of love" because it was converted to a "monotheon."

The Romans had mastered the principles of arches, and a dome is an arch spun around a vertical axis. Continuous floor space as vast as the Pantheon's was possible so long ago only with the structural supremacy of the sphere. Roman arches and domes were basically compressional structures: stacks of units of compression–resistant materials, such as stone, brick, and unreinforced concrete. That the Pantheon pushed the envelope of possibility set by such materials is shown by other heroic efforts in subsequent domes. Brunelleschi's dome in Florence and that of St. Peter's in Rome reached limits at about the same size as the Pantheon. Back another order of magnitude in time (and down in size), clans in the late Paleolithic Ukraine piled mammoth tusks and mandibles into sturdy domes that bore the brunt of Ice Age winters.

A pile of domes. *From top:* towhee bird egg, pistachio, beetle, buckyball molecule, and tortoise. Under the tortoise, not shown, it's more domes all the way down.

Dome of the gods: the ancient Roman Pantheon.
PHOTOGRAPH BY LYN HUGHES.

But human-made domes are not usually omnidirectional in orientation. It is here that nature excels. Our own skulls, for example, are not dependent on orientation for strength. We can stand on our heads. But don't turn the Pantheon upside-down. The various domed shells of nature need to be stable in any direction—an egg rolling in a nest, a beetle tumbling onto its back, a nut falling to earth. Many natural types of tough, fibrous materials offer abundant resources for weaving a continuity that can withstand both the pull of tension and the push of compression.

Bucky Fuller's geodesic designs, too, require materials strong in tension as well as compression. For my own hex-pent studio dome built after architecture school, I chose wood for the lattice frame, and succeeded in clear-spanning twenty feet with a total structure of only two hundred pounds. Fuller's geodesic designs are thus close kin to nature's own. In the icosahedral shell of a virus, the atomic-bonded lattice resists both stretching and compressing.

Domes show a tight concurrence of two key physical properties of spheres: the strongest structure for limited material and the largest volume. This two-for-one (at least) deal has not gone unnoticed by the evolutionary powers of nature. But it may be only the knife of reasoning that splits the biological attributes of spheres into strength, capacity, and omnidirectionality. A round seed with crush-resistant shell simultaneously packs the maximum mass of plant embryo, resists the mammal's taste-testing, and rolls easily when pushed by an ant. (Underground the ants consume the edible part of the outer rind, and, if lucky, the seed is rewarded with a superior place to germinate.)

Various aspects of the sphere, in the seed's case, function at different stages in the life cycle. Yet they are all wrapped into one package. The roundness of oranges is obviously a case of minimized surface, but roundness is also useful for deflecting from any direction the wind's potentially ripping gusts.

Overall, thinking about spheres can carry a quester to multiple levels of understanding. Within the grape interplay expansion and control. The resulting shape has minimal surface area, and this, in turn, has vital functions. Then too, vitality affects survival, thus bringing to bear evolution's ability to change the enzyme loops and cell cycles within—completing the cycle of logic around the sphere from its internal dynamics, to the environment's pressures on design, and back again. These levels are related (but not in one-to-one correspondence) to the conceptual sphericity given by a full sweep of question words—what, how, where, when, and why—and Aristotle's levels of causation—material, efficient, formal, final.

The metapattern of sphere arises in and connects the various physical and sometimes ecological and evolutionary levels within a particular example. It also can help us to see relationships and correspondences between wildly different phenomena, as in the grape-moon koan. But a well-posed koan can also carry the seeker into realms of the psyche.

 ## Spherituality

Enter the spheres of myth, inspiration, and understanding. The intoxication of sweet grapes and full moons calls forth the divine: Bacchus, Demeter, Astarte, Orpheus, Dionysus, Krishna. From Brunelleschi's miraculous dome arose the glory of Florence and the quest for creations still greater. Today the buckyball is proclaimed "magic molecule" and "Molecule of the Year."

A ball of fibers can be a focus, able to rally the masses to frenzy and catharsis. Roots of ball rituals run deep. Black Elk of the Oglala Sioux described the meaning of one of the seven sacred rituals, Tapa Wanka Yap, a ball game: "The game as it is played today represents the course of life, which should be spent in trying to get the ball, for the ball represents Wakan-Tanka, or the universe. . . . In the game today it is very difficult to get the ball, for the odds—which represent ignorance—are against you."

Universe, divinity, grandeur, catharsis—the deepest facets of sphericity are many and complex. One such facet pertains to mythic explanations for the mystery of the origin of the world and its forces. Commonly, the current universe began with a crack in the cosmic egg. Various Neolithic peoples depicted the cosmic egg within birdlike goddesses. Many deities worldwide—Phanes and Eros of the ancient Greeks and Nargaruna of the ancient Finns—burst from eggs.

Because real-world evidence of fertility so often comes in the form of spheres—from snake eggs to seeds—a spherical motif for the mythological birth of the universe was perhaps inevitable. The crafters of myth, however, also placed this shape in diverse sites of reverence. Hindus sculpted perfect spheres for breasts, Christians domed baptism sites, and Zen monks painted circular ensos for the enlightened con-

Inside the beehive tomb of Atreus. Author stands inside a Mycenaean tomb, dreaming of a final theory, of the gestation of babies and laws, of Neanderthal burials, of the baptistry in Pisa and one of the five domes of San Marco in Venice, of Hindu goddesses, of a stone finial atop a gatepost of a mason in Slovenia—of birth, death, sky, god, equality, and perfection.

sciousness of satori, or rebirth. Today in Washington, D.C., the great egg crowning Capitol Hill proclaims where laws are bred and born.

A dying amoeba may shrivel to a sphere, but metaphorical and mythical use of the sphere to depict or explain death is not so obviously a transfer from biology to mythology. Nevertheless, as ant hills show, piles of earth repose naturally rounded. Such least-effort shaping perhaps accounts for round mortuary mounds in Neolithic Holland and pre-Roman Etruria. Structural efficiency would have played a part in the underground rounded or somewhat conical dome tombs of Mycenae and of other cultures of the late Megalithic. But by the time of the powerful Ashoka, who blanketed Asia with Buddhism more than two thousand years ago, the sphere as preferred icon for the revered dead was highly refined; for example, the Buddhist stupas that housed religious relics each sported a massive central dome called the anda, or egg. Imperial Rome had its domed tombs of Augustus and Hadrian; Islamic tomb builders, too, made the dome a dominant feature. The early Christians erected domed martyria to house and worship body parts and other memorabilia of famous martyrs.

The spherical mode of tombs and other tributes to the dead may combine structural with symbolic functions—notably the dome as sky. Analogous to their lives under the sky, the political or religious elites could continue after death under an overarching roundness all their own.

The sky dome, overall, manifests a tremendous draw on the human spirit. Ancient Persian rulers decorated their domed "heavenly" audience tents. Numerous rulers of the Near East and Mediterranean were fond of the ambiance afforded by domelike canopies, or baldachins. And so came the painted pantheons of gods within Roman domes and the Christian "domes of heaven," with Jesus or the Father, in person, surrounded by heavenly hosts up high in the round.

The sky is a major (for many cultures, *the* major) god realm, the place of the untouchable master of cycles. Sky can even *be* god. Thus birth, death, sky, and god are all kin with the sphere. And out of these arises a final association: sphere as power. Both sky and god have power, power over birth and death; and rulers (the stupa-builder Ashoka and Pantheon-builder Hadrian) have power to give birth to edicts controlling life and death.

One way to show power is to hold it. Hand-held spheres of power have been symbols for many cultures. Jesus has often been depicted holding a sphere, as have the emperors of Rome. The sphere as power was reflected in the wish-granting jewel in the palm of a bodhisattva. It was the orb of power brandished by European kings and queens, the rotundum revered by alchemists. It is still secured under the paws of sculpted lions from Florence to St. Petersburg.

Political or cosmic rulers have the whole world in their hands. The self-assured faces of the holders of spheres make it seem they were

Spheres in the hands and heads of the mythic imagination. Christian Madonna and Child; heads from Buddhism, Australian animism, and Christianity; Buddhist goddess holding the wish-granting jewel; Cleopatra; Florentine lion in the Piazza della Signoria.

born for the position. But all hands are so endowed, perhaps even evolved, to comfortably palm a rock, or an apple.

Another ubiquitous position for spheres is around heads. Perhaps this magnifies the rotund, hard fact of our skulls, building an iconographic bridge between the mind beneath the bone and the body beneath the sky, between the births of ideas from brains and organisms from eggs. The round heads of African Ashanti fertility icons suggest such a concatenation of birth symbols. To my eye, the solar disk surmounting the heads of painted and sculpted ancient Egyptians descended in time and space to actually surround the heads of Jesus the Christ and Siddhartha the Buddha, and their respective saints and bodhisattvas. Both Buddhism and Christianity emphasize a rebirth of self with omnidirectional compassion, power of serenity, and guidance for facing death—all themes of the sphere.

In European traditions a sphere around a head has a political manifestation: the crown. Wearing these metallic halos, rulers displayed their responsibilities. Here is one theoretician (or propagandist), speaking in 1610 about James I of England: "The Sphear-like forme of his Crowne doth denote the even roundnesse wherein he proceedeth to every one, as well towards the smal as the great, the poore as well as the rich . . . like unto the Geometrical point, which beholdeth all his circumference in one and the same proportion."

Equanimity, then, is yet another aspect of spherituality, showing the subtle transfer of meaning between geometry and the evolution of consciousness. How fitting that the sphere inspires behavioral ideals, since it is itself an idealization, approached yet never fully achieved. With the sphere we have a concrete shape to model our values. Consider: in the optimism following the French Revolution, many architects drew designs for Temples of Equality, some intended to be fully spherical spaces.

As the physical attributes of spheres—omnidirectionality, surface area, and strength—are often separable but not separate, so the sphere's archetypal attributes—god, power, equanimity, idealization, perfection—seem to merge. A stunning example of the mythic connections between sphericity, divinity, and perfection comes from Plato's description of the origins of the universe, from his *Timaeus*:

> Creator compounded the world . . . as far as possible a perfect
> whole and of perfect parts . . . leaving no remnants out of which

another such world might be created . . . that figure . . . which comprehends within itself all other figures . . . the form of a globe, round as from a lathe, having its extremes in every direction equidistant from the center, the most perfect and the most like itself of all figures . . . the surface smooth all around . . . because the living being had no need of eyes when there was nothing remaining outside to be seen.

The same imagery reverberated through Aristotle, who supplemented Plato's logic with astronomical detail. Ptolemy of Alexandria continued the trend and, with authority that reverberated for one and a half millennia, unfolded the mathematics of sphericity into an all-encompassing explanation for heavenly motions. The sphere became the paradigmatic tool of cosmologists.

Then came Copernicus. In what we now regard as the prototypical scientific revolution, Copernicus wrote his great work on astronomy—a study he believed to be "head of all the liberal arts" because its subject is "the godlike circular movements of the world." Copernicus begins his *On the Revolutions of the Celestial Spheres* with this argument:

First of all, we must note that the universe is spherical. The reason is either that, of all forms, the sphere is the most perfect, needing no joint and being a complete whole; or that it is the most capacious of figures, best suited to enclose and retain all things; or even that all the separate parts of the universe, I mean the sun, moon, planets, and stars, are seen to be this shape; or that wholes strive to be circumscribed by this boundary, as is apparent in drops of water and other fluid bodies when they seek to be self-contained.

Copernicus's scientific successor, Kepler, followed this tradition by noting that the sphere is the "image of God the Creator and the Archetype of the world." Nevertheless, Kepler put the first dent in the archetype. He cracked the perfection of planetary spheres (circles) with the discovery of elliptical orbits. The end of the world's "godlike circular movements" ushered in the objectivity so essential to science. A conclusion drawn from sound observations and calculations, no matter how distasteful, could not be willed away.

No longer archetype of the cosmos, the sphere persisted as tool of science. Galileo single-handedly invented experimental physics by rolling and dropping balls. In studying the rainbow, Descartes traced light paths within perfect spheres of water. To compute predictions using gravity as a force, Newton first had to prove that a sphere could be mathematically replaced with a central point of equivalent mass. Coulomb electrified various material balls, and von Guericke created vacuums within spheres in his sensational horse-tugging demonstrations of air pressure. Einstein idealized the microscopic jitters of

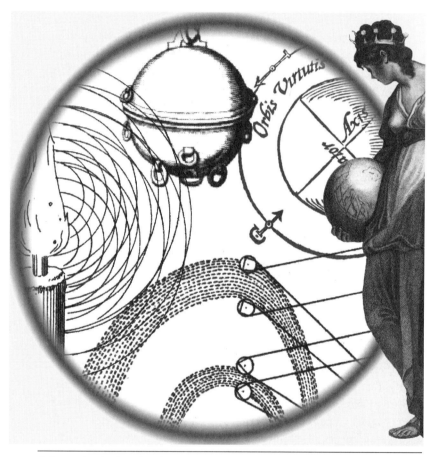

Spheres in the hands of science. The globe-holding muse of science, from an allegorical picture by Corbould, surveys a Cartesian analysis of rainbow droplets, Huygens's wave theory of light, Otto von Guericke's vacuum sphere (which resisted the tug of sixteen horses), and a diagram of a spherical lodestone of Gilbert, with its *orbis virtutis* or "sphere of activity."

Brownian motion as bouncing spheres in the first calculation of molecular diameters. Alternate systems of geometry to the Euclidean plane were discovered upon spherical surfaces. And why is that mathematical miracle, the zero, round? One could argue that much of science has grown from the early attempts to explain data using the powerful simplifications possible with the theoretical sphere.

The trance of sphericity has penetrated beyond myth, beyond early science, into everyday language. Why do we say "circle of friends" rather than "square"? Why do we admire those who are well rounded, not well triangulated? Why do we speak of spheres (but not icosahedra) of influence, knowledge, and power?

The image is about encompassing, about including some parts of the universe and excluding others. The sphere is the elementary division of inside and outside. If there were to be a symbol for the "thingness" of anything, I would guess we'd agree on the sphere, or, on the page, the circle. That is indeed what I will use in this book to visually represent entity, existence, being.

Then too, my goal is to present patterns spherically, that is, across disciplines, spreading into all directions of knowledge. And metapatterns can be thought of as those key patterns that exemplify the sphericity of the deepest forms of knowledge, the interlinking of all things and ideas in the universe.

Where today do we stand regarding the themes of spherituality? It seems that much of the sphere's archetypal power has vanished. Our leaders no longer hold orbs. Our heroes no longer have halos. The urban masses rarely revel in the full celestial starry dome, and astronomers know the music of the spheres to be chaotic discords. The temples of equality were never built. And god is not an ideal, sentient shape.

Where do we stand? I suggest looking literally beneath our feet. Despite its near disappearance from the themes of art and architecture, the sphere does live on, concentrating all its profound power in a single image: Earth. Earth is our birth, our death, enveloping us within its component spheres of biosphere, atmosphere, hydrosphere, lithosphere, technosphere, and noosphere. Planet Earth becomes our god, inspiration, truth, perfection, equality, and source of power. In hand-sized pictures, we each can hold our planet, an icon as powerful as royalty's crown or divinity's halo: Earth from space. Our fate and the fate of this sphere are one. Thus inspired, Maurice Strong, Secretary General of the 1992 United Nations Conference on Environment and Development, proclaimed: "We must now forge a new 'Earth Ethic' which will inspire all peoples and nations to join in a new global partnership. . . . Earth is the only home we have; its fate is literally 'in our hands'."

According to Black Elk, during the sacred ritual of Tapa Wanka Yap, after the ball has been tossed sequentially to the west, north, east, south, and above, it is returned to the girl who that day is Rattling Hail Woman. As she holds it, all begin to chant:

> Behold today Rattling Hail Woman, who holds in her hand a ball which is the earth. She holds that which will bring strength to the generations to come who will inherit the earth; and the steps that they take will be firm, and they will be free from the darkness of ignorance. Rattling Hail Woman stands here holding the world, and, from this day on, this ball will belong to generations to come, and they will rejoice as they walk hand-in-hand with their children.

Shape's spectrum in the leaf of a gamble oak. A small species of wasp lays its eggs in the nutritious leaf veins, thus inducing the oak to produce bulbous swellings called galls that serve as homes for the larvae.

Sheets and
Tubes

Shape's Spectrum

 As a prism splits light, the mind splits shape. A prism spreads white light into a spectrum of colors, sorted by wavelength (frequency). By what attributes can shape be sorted? The choices are vast: degree of symmetry, complexity, fractal dimensionality, even aesthetic harmony are just a few examples. But given the sphere's unique set of qualities, I find it fruitful to posit a spectrum according to degree of sphericity.

The spectrum of color in visible light, from red to blue, is a gradation of wavelength from long to short, of frequency from low to high. Along what spectrum does sphericity grade? Contrasted to the sphere's omnidirectionality would be shapes with direction. Because the sphere disperses forces, its spectral opposite might channel them, and minimized surface areas would grade to those maximized.

Two broad classes of shape meet these criteria to serve as counterpoints to spheres. One can be imaged as a pancake, the other as a spaghetti noodle. To make the first, squash a sphere, like flattening a lump of dough into pie crust. To make the other, stretch a sphere, like pulling taffy. In mathematical terms these counterpoint shapes are planes and lines. But because my concepts are not the idealized forms of mathematics, rather, functioning three-dimensional objects, I will call them sheets and tubes.

Functions of Flatness

Cooking a pancake begins by plopping a spoonful of batter into a hot pan. Outward across the metal plane the batter spreads, away from sphericity. We live between floors and ceilings, we sleep between sheets; our households are imbued with horizontals. Outside, horizontal fields of clouds billow upward from the flat bottoms that form at the lifting condensation level, the otherwise invisible transitional surface where water vapor in lofted currents of air chills into droplets. The oceans—ultra thin pancakes of water—also repose as surfaces perpendicular to the line of gravity's pull.

Physical, biological, and cultural sheets. Conforming to the plane of gravity are clouds at the lifting condensation level. The floor of the Gran Trianon at Versailles, a rock-encrusting lichen, and the veined wings of a dragonfly echo the sheet motif of a leaf. The New Guinean musician has apparently floated into the palace on his river raft.

During a research voyage across the Atlantic, I was able to taste near-freezing polar brine. This was odd because the ship, gently rocking upon the endless plain of the air-sea interface, was near the equator. We had successfully fished for water at discrete depths using a high-tech sampling rig. Just several miles below (reachable in an hour of cautious winch-time) were the deep layers of cold water that had not contacted air for hundreds of years—not since they had cooled and sunk in a polar sea and begun the journey sliding far beneath the surface. That water masses travel as globe-stretching, interleaved sheets has important ecological consequences for the cycling of nutrients (to cite just one aspect). Specifically, these benthic cold layers supply the oxygen needed to recycle the organic debris that falls from the tropical surface into dissolved nutrients bound back to the surface.

Proportioned like slightly curved postcards, gyres of air and water transfer matter and energy from place to place across vast distances. All the geophysical sheets (and sheets within the sheets) of air, water, rock, and ice present extensive interfaces. Together these geophysical sheets are only an infinitesimal speck of Earth's mass, yet their total surfaces are many times greater than the simple geometric surface of the great sphere.

One multiplier of transfer surface is not a geophysical sheet but a biological one. The green leaves of all terrestrial plants sum to an area about four times that of the continents. Leaves are oriented and shaped for a biological function: to absorb light.

One way to grasp how well leaves are designed to collect light is to compute their equivalent volume spheres. Like folding a pie crust back into a ball, convert the leaf to a sphere of the same volume. Then compare the surface areas. Typical leaves I have measured have surface area enhancements of twenty to forty times. That is far greater than what the typical pancake achieves. Such relations between area and volume have pulled living forms, as they evolved, to particular places in shape's spectrum. And so the grape vine puts out its fruit as spheres and its leaves as sheets. Were we active creatures to photosynthesize our own metabolic requirements we would need about twenty more times the amount of skin than evolution has presented us with. That would make getting in and out of cars quite difficult.

Another type of energy transfer is accomplished by sheets that are designed to capture motion. Sails, turbine blades, and ear drums catch and transmit the power of pressure-driven gases. Wings of butterflies, birds, and bats, the tails of fish and the flukes of whales reverse the direction of power—which now comes from the sheets themselves rather than the surrounding fluid. Webbed frog feet, propellers, and the vibrating sheets of speaker cones do the same. The sheet is active, not passive.

Sheets can also transfer matter. Again, the supreme example is the leaf. Across its extended surface enters the plant's prime building material, carbon dioxide. Across the same surface exits the by-product of photosynthesis, oxygen. Leaves also transpire water, akin to our sweating. Transpiration not only cools the leaf; it is the mode by which nutrients are pulled upward from the soil to where they will be used. In aquatic animals, gill surfaces reverse these flows of oxygen and car-

Sheets for transferring matter, energy, and messages. For energy: solar panels (author's head for scale). For matter: delicate lamellae on the gills of a rainbow trout (about 50 microns apart). For messages: carved glyphs on a Mayan slab stela. The tail feather of a wild turkey (here in detail) transfers both energy (forces to the air to stabilize flight) and messages (here to hide information—for camouflage).

bon dioxide. The microscopic sites where these gases are exchanged during respiration (in plants as well as animals) and by photosynthesis (in plants alone) are, respectively, the inner membranes of mitochondria (convoluted like our brain) and the stacked thylacoid disks of chloroplasts. Inside these organelles inside cells, across pleated membrane surfaces move the materials that power life.

Energy, as well as matter, is sometimes the aim of mass transfer undertaken by living organisms. Oxygen is both an essential ingredient of biological molecules and, as a pure gas, a carrier of chemical potential; the fly captured by a sticky web supplies both matter and energy for the spider's metabolism. And both mass and energy can be carriers of a third item transferred across sheets and which is essential to life: messages. Light—given meaning in reflected emission from flower petals, butterfly wings, Mayan stelae, billboards, flags, and the sheets in our billfolds—is a universal bearer of what Gregory Bateson termed "differences that make a difference."

Sheets allow the display of the most message for a given mass. Pages are sometimes called "leaves." I have computed that a page from *Science* magazine has a surface enhancement of two hundred times over its equivalent volume sphere, bettering sizzling batter and even green leaves.

Thus shapes can be sorted along a gradient of flatness—a spectrum that has spheres at one end and the thinnest sheets, such as oceans and science journal pages, at the other. Sheets contrast with spheres in their surface-to-volume ratios, and also in their directionality. Hurling a frisbee is quite different from throwing a baseball. Yet sheets are not unique in this contrast.

Tubes for Transport

Tubes, too, have greater surface areas than spheres of equivalent volumes. A reed of No. 25 spaghettini from my pantry has between four and five times more surface area than its equal volume sphere. Compared to leaves, that does not seem a great achievement, but it is still something, and (as I will explain) tubes excel in other ways.

But first consider the tube's enhancement of surface to volume. Thinner noodles cook faster because of greater surface enhancement. One unforgettable lesson in life is to not walk away from a pot of boiling angel hair pasta.

Thus where design calls for transfer surfaces, tubes as well as leaves can be a solution. A pine needle as leaf, for example, captures photons and exchanges gases with more surface than if its biomass were bundled into a ball. The pine tree, moreover, excels in surviving drought and cold; it is by no means an evolutionary failure. Linearity enhances the surface areas of intestinal villi, tapeworms, root hairs, our body's sixteen billion capillaries, and the white threads of fungal hyphae in soil detritus. Like sheets, tubes also exchange momentum with their environment. The tail of a sperm and the hairlike cilia of cells are there to propel. By contrast, the tiny spines projecting into cochlear fluid work like tuning forks in reverse, absorbing wave patterns inside the mammalian ear.

Yet, as might be expected from their unique topologies—linear stretch versus planar squash—tubes and sheets are not generally interchangeable in places that need surface. We fly a kite as a sheet attached to a tube, not the other way around. We paddle a canoe holding the oar's tube with its sheet in the water. Where the press of maximizing area overwhelmingly drives the design, sheets will dominate. There-

Tubes as transfer surfaces. Not all leaves are sheets, as seen here in a pine. Not all blimps are Goodyear, as seen here in the single-celled microorganism *Tetrahymena*, whose tubelike cilia serve for propulsion.

fore, when tubes appear in biology or technology, other attributes of the tube probably have pushed the design in that direction. What are these?

Viewed on end, a tube looks tiny. This minimized surface along an orientation is a second major attribute of tubes. The threadlike forms of fungal hyphae give them enhanced total surface for absorption and, simultaneously, reduced end surface for extension during growth. Overall, streamlined and elongated shapes facilitate smooth forward motion through air, water, soil, and other matrices of obstruction. Consider blue fin tuna, earthworms, Boeing 747s, ships, footballs, forks, stamen tubes, carpentry nails, arrows, rockets, whales, canine teeth, cars, roots, and river otters.

Because they will encounter some degree of resistance, all forward-moving tubes in life and technology need a measure of rigidity as structural columns, to transfer the forces encountered from the front toward the back. This ability to transfer forces along lines is a third major function of tubes, and, for many, as structural tubes, their raison d'être.

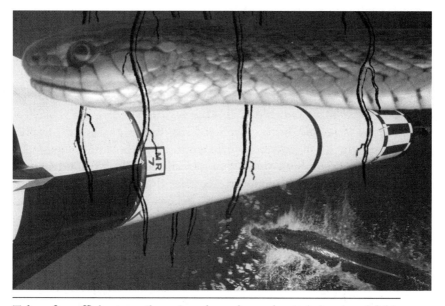

Tubes for efficient motion. A snake and a rocket, a swimming dolphin and growing roots all use the tube metapattern to minimize surface area for a leading end.

Generally we classify structural tubes by the types of forces they accommodate. Kite strings, spider web lines, and the guy wires inside cells (called microfilaments) are all good at handling tension. The Greek Parthenon's marble columns and the tent posts inside cells (called microtubules) are good at handling compression. Sometimes forces are actively transferred by tubular structures: a rope around a pulley lifts a bucket, a car's driveshaft powers the wheels. In other cases the transfer is more static, as along the string between kite and hand, or in the Parthenon's columns from pediment to ground.

Life presents a plethora of structural tubes, from the largest of organisms down to molecules. The tubes of plants—the redwood spire, the gangly bush, the slender grass, the mite-sized stalks of moss—all have ancestry in the tentative experiments that, several hundred million years ago, first pushed cells off the surface, separating earth and sky. A key invention was cellulose, that long, fettucine-like array of carbon and hydrogen atoms. Collections of cellulose molecules form the fibers of plant cell walls; these in turn are embedded within tubular conducting vessels called tracheids, which are aggregated into the large strands of fibers visible when we cut wood. A tree trunk is thus an excellent object for contemplating levels of structural tubes. A similar telescoping system of structural tubes echoes within animal muscles: from fibers, to smaller myofibrils, still smaller filaments, and, on the molecular level, lengths of actin and myosin—the pair of molecules that ratchet along each other during muscle contraction.

Stephen Wainwright, a biologist at Duke University, has explored these issues in a wonderful book, *Axis and Circumference: The Cylindrical Shape of Plants and Animals*. Evolution has bestowed cylinders (his preferred term) on all scales from polymers to whales because of their functional advantages. Structurally, then, they are inevitable. Wainwright outlined the probable evolutionary sequence by order of the functions fulfilled by cylinders: beginning with the cell-bridging abilities of long and spindly molecules, such as mucus; then preferred parallel orientations of polymers as the basis for aligning cells; then fast axial growth rates leading to the filaments and tiny worms of early plants and animals; and finally three major body strategies—branched cylinders (plants, corals, bryozoans), fiber-wrapped hydrostats (worms, squids, tongues), and kinetic frameworks (brittle stars, arthropods, vertebrates).

Perhaps the functions of all these biological structural tubes can be summed up in a word: reach. Whether for resources, lift, movement, or the exchange of gametes, "the most efficient use of materials in support systems that reach out," according to Wainwright, "occurs in cylindrical bodies." Tubes clearly beat spheres because "to outreach a neighbor by growing spherically is expensive."

The reach of structural tubes goes beyond biology—back into physics and forward into technology. A universal property or function of tubes is that they maintain distances. From tubular bonds between atoms to the beams of buildings, tubes offer escape from inward-turned sphericity. Together, the lines of tension and compression provide separations at particular distances to create forms in space. The

Structural tubes. Framework of the Eiffel Tower and tendrils of a grape vine. Usually a grape vine seeks a tree or trellis for support, but this one apparently suffers from megalomania.

soaring levels of tubes as struts in the Eiffel Tower provide an architectural echo of the structural levels within trees.

We have seen that the tube as structural connection allows for a transmission of forces. This hints at the most general—and probably most important—property of tubes: providing a passage for flows between one place and another. As in the case of sheets, the flows may be classified (often usefully, but not always unambiguously) as mass, energy, and messages.

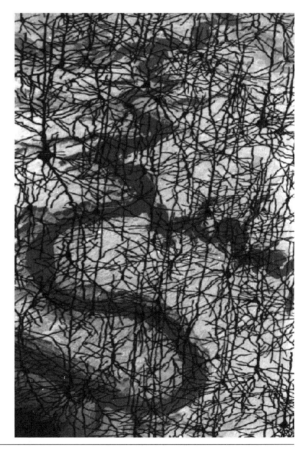

Tubes for transport. A river with its tributaries in the western United States is here viewed through a network of nerves channeling signals in a human brain.

The flows of vital fluids within tubes in similar-looking net-works—rivers, plants, blood systems—can lead to a view of Earth itself as an organism. In channels water moves, as directed by gravity or the governor of California. We build canals, and we call the passageways in sponges same. A cell's microtubules guide the movements of its materials just as highways and railroads do for us. Our throats and bronchial tubes are conduits for gas exchange; steel pipelines shunt natural gas from fields and tanks to dispersed networks in cities. Along wires (again, a solid tube) electricity flows and carries power, or, if modulated, messages. To transport lines of speech and print in pulses of laser light, we lay fiber optic cables beneath the Atlantic.

In these examples of tubes for transport the primordial function of the distance-bridging structural tube is enhanced. Tubes provide paths for long distance transport with least material. Places are connected, but the integrity of each is maintained. This feature has profound significance for the role of tubes in the maintenance of what we normally think of as relations between objects in space. And consider that tubes can be found within sheets and within spheres (veins of a leaf or grape, structures of grids within feathers and airplane wings, masts and ropes of sails).

The various functions of tubes—conceptually separable but not really separate—often concur. Leaf veins provide structural support as well as transport systems for water, sugar, and ions; roots anchor a tree as they probe long distances with minimal material and small frontal area; large blood vessels (veins and arteries) serve primarily to transport, while small blood vessels (capillaries) specialize in transfer across their enhanced surfaces. In shape's spectrum, opposite to the sphere's omnidirectionality, tubes are the epitome of directionality. So ubiquitous for so many systems are the directional properties of tubes that I have come to wonder: Are the tube and sphere an archetypal pair? To test this idea, I have looked for expression of the tube in myth and metaphor.

Synergies of Things and Relations

People, like animals, traverse wide stretches of land-scape between points of resource concentrations. Repeated migrations can make these paths well worn, or at least well known. People may even make diagrams of such networks of nodes

and paths. The aboriginal artists of Australia have refined such diagrams to a high art, as well as a sacred knowing. Their networks of
concentric roundels and connecting tubes, often much like the geometric patterns of Fuller's geodesic domes, are road maps for both the
physical and the metaphysical. In their metaphysical aspects the
roundels are dreamings, great events in the mythic past, where, for
instance, Wichetty grub emerged. Both levels, the physical and the
mythic, are served by the same pattern: circles linked by lines.

A similar pattern of circles and lines is used by the Jewish mystics,
the kabbalists, to portray the Tree of Life or Tree of Holy Fruit. The
circles—cosmic *things*—depict the ten Sefirot or Divine Emanations,
the attributes, utterances, powers, vessels, faces, hands, or garments of
God. The lines—cosmic *relations* or paths—correspond in one interpretation to the letters of the Hebrew alphabet.

The same concept—a tube-connected system of spheres—is used
by the American Society of Agronomy. The ASA sees itself as one bubble connected to another tier of bubbles representing issues tackled by

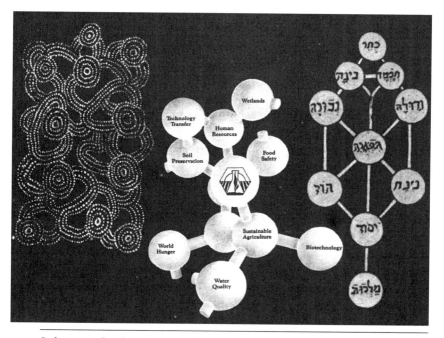

Spheres and tubes as mental models. Examples here: a bark painting of an
Australian aboriginal "dreaming," a diagram on a poster from the American
Society of Agronomy, and the mystical geometry of the Kabbalah.

its members' advanced science: world hunger, biotechnology, soil preservation. Not quite divine utterances or dreamings, this system is nevertheless mythically crucial to the ASA.

The aborigines, kabbalists, and agronomists, despite their diverse goals in diagraming big chunks of their respective universes, have converged to strikingly similar patterns. To be sure, the diagrams differ in many details—for example, whether or not they are symmetric, labeled, or expandable. But the base pattern that underlies their creation is nature's Tinkertoy: the world as things (spheres) connected by relations (tubes). Here the tube takes on mythic aspects, as partner with the sphere, in visualizable lines between concepts in our minds. What gives it so much power in this partnership?

Consider cases in which a physical object consists, in essence, of a single sphere and a single tube. The amoebas of slime molds, when aggregating to produce spores, build a stalk topped with a sphere. In this irreducibly simple system the sphere and tube have clear roles: the bulky sphere for the formation, storage, and release of spores, the tube for support away from the surface. Our water towers look similar: the sphere stores the water, the tube provides support and a system of pipes for the up and down flows. A sperm's round head contains the DNA, its tail propels. The sphere and tube in such systems are functional complements. Usually the sphere is storage and site for transformations; the tube supports, propels, and supplies the sphere. Two ends of shape's spectrum, their properties are functional complements.

Humans have elevated this functional synergy to an aesthetic synergy. Sacred sites of architecture give visual proof of this archetypal pair. Dynamically poised adjacent to the cathedral with Brunelleschi's dome, and completed even earlier, is Giotto's tower. This tower, the campanile, stimulated a Florentine pride and vigor known as campanilissmo. In Katmandu, a Tibetan Buddhist stupa embodies earth and heaven as the egg topped with a spire. Domed Islamic mosques also have jutting towers, the minarets. Juxtaposed to the dome in Washington, D.C., at the opposite end of shape's spectrum and the reflecting pool, rises a white stone obelisk.

On more personal scales, the pairing also abounds in painted and sculpted icons. In engravings, Queen Elizabeth I cradles the royal pair of scepter and orb. Popular in European court portraits (sometimes with sword instead of scepter), this synergy of power symbolism, with the omnidirectionality of fairness and directionality of action, attends to both the spheritual and the spearitual.

Other cultures, too, have imaged this pair: Egyptians, native Americans, Coptic Christians. What are the roots of such symbolism? Many possibilities come to mind: digging stick and gourd, big stick and big rock, sword and opponent's head, bat and ball, even the flute and humpback of the shamanic Kokopelli of the vanished Anasazi. I conjecture that the sphere and tube mythic pair emanates from an at least intuitive recognition of the two poles of shape's spectrum, an elemental synergy of form and function that cultures could not help but incorporate into religious and political iconography.

Royal and religious tubes might have developed as icons of paths to other realities. Consider the shaman's path of transcendence, the ladder to heaven, the *axis mundi*, the bridge from ordinary consciousness. The sphere and tube are thus god and the way to god.

Sphere and tube as archetypal pair in art and architecture. The campanile and duomo—tower and dome—rise in the heart of Florence. Using her domelike garb for flotation, Queen Elizabeth holds the scepter and orb, symbols of royalty.

In feudal Japan, helper gods would arrive along a pole planted in the ground during battles. During Sioux rituals, with bowl full of "all objects of creation and the six directions of space," the sacred pipe was leaned vertically against a rack—the pipe's stem was the path, like lightning, the axis joining earth to heaven. Vertical tubes mark sites of revelation and culmination, like the bodhi tree of Buddha's final enlightenment and the cross of Jesus. Tubes take us to exalted spheres.

Together the sphere and tube as functional, aesthetic, and cosmic synergy may supply part of the power for the coexistence of snakes and eggs in mythology. Phanes, the Greek god of light, bursts from the cosmic egg entwined by a snake. Around 300 C.E. an Ohio River culture labored to shape a giant snake and egg of earth. Eve was urged by the snake to partake of the sphere of knowledge; the snake was the go-between, the bridge taking Eve to the fruit, which transformed history.

Combinations of tubes and spheres in myth, art, and architecture may stem from a primordial visual gestalt. Very young children (nursery school ages) will draw a person as a big round body-head with radiating lines for arms and legs. So universally does this figure emerge from children as a stage in their visual fluency that child psychologists have named it: the tadpole figure. Is this how children not only portray but conceptualize themselves—as a big volume of being reaching out? Is the tadpole figure the primal body image we all carry within?

If so, the image serves us well as a model for the way things work. Not just us, but many other beings radiate from central spheres, reaching out with thin extensions to touch, grab, give, receive, and connect. Tubes bridge the spaces between beings, and in bridging, create space itself—not only the space of physics with Tinkertoy atomic bonds but the complex dimensions of functional spaces for life and culture. By way of tubes, the spheres of the universe hold hands and dance.

Even more, the tadpole figure, I surmise, has moved from root image for our bodies to rich symbol for our minds. As scientists and scholars we then project the tube back again into the physical realm. We seek to reveal the relations of nature, not merely those present in tangible tubes, but those "far more deeply interfused." And with the help of tubes extended from eyeballs, Galileo gained a better view of the big spheres, van Leeuwenhoek the small. In Santa Croce church of Florence, a sculpture tells all. There a marble Galileo peers out; the new royalty of scientific power cradles in one hand not a scepter but a

Tubes and spheres in mind and nature. *Clockwise from bottom:* prehistoric rock art in New Mexico; water tower; child's drawing; penstemon seed case and stalk; adult's sketch in the game called "Pictionary" (here the word to be graphically portrayed was *child*); eyeball with muscles; living foraminiferan; alchemical vessel; tadpole of a canyon tree frog; child's drawing; fruiting stalk of a fungus; child's large watercolor.

telescope and in the other not the royal orb of power but a sphere seemingly plucked from orbit.

More abstractly, researchers connect tubes of behavior to their objects of study. Experimental manipulation emplaces tubes as probes—instruments, protocols, theoretical questions. Back through the tubes of observation flow data, also enhanced by technologies and systems of theory. The goal is to reveal things and their deeply interfused, radiating relations: fields in physics, food webs and energy flows in ecology, money exchanges in economics, ocean oxygen fluxes in earth system science, crop responses to sunlight in agronomy, and on to a near-infinity of human concerns. We connect our own minds by behavioral tubes to enliven, discover, remake, appreciate, and contemplate the spheres and tubes we see at all levels, in all endeavors. As shown by the aborigines, kabbalists, and agronomists, this is the way things are—or at least the way we think.

We in earth system science, my own field, examine interactions among the biota, atmosphere, oceans, soils, and rocks. The scientific synergy of spheres and tubes as a tool for discovery is alive and well in both theory and practice. And I am impressed with it. Through mathematical modeling I arrange and rearrange nature's Tinkertoy to decipher the workings of the global carbon cycle.

Let's diagram how one might begin a detailed mathematical model for Earth's carbon cycle. A circle in the center represents the reservoir of carbon (as carbon dioxide) in the atmosphere. Around it three other circles represent reservoirs of carbon in underground fossil fuels, land plants and soils, and the global ocean. Lines between each of the three and the atmosphere represent fluxes of carbon by burning, photosynthesis, respiration, gas diffusion, and so forth. Although highly simplified, the model's components are workably analogous to those of Earth.

Perhaps we wish to compute how the contents of the four major reservoirs might change over time. Because the flow rates change, too, imagine valves on the tubes that allow more or less flow. But what controls the valves? For this we need circles of another type. Not physical analogs like the reservoir circles, these new things—the control parameters—include items like ocean area, wind-related gas exchange rates between the sheets of atmosphere and ocean, forest growth and deforestation rates, and the crucial factor of human demand for nonrenewable energy. To show their effects, we connect these control circles to the valves on the flow pipes.

In our homes, for example, and in many technological systems, both types of conceptual lines are indeed present as physical lines. Gas flows in a pipe to the furnace, which receives control signals passed in a wire from the thermostat. The control "things" in the carbon cycle, while not as tangible and tunable as thermostats, are just as real; they can even be affected by one another and by the levels of the reservoirs. For example, the atmosphere's carbon dioxide level will affect the growth rate of plants. It does so directly and also by driving rainfall and temperature changes. When actual or potential impacts of human dumping of greenhouse gases into the atmosphere disturbs enough people, we may be able to add a new line from the atmosphere to the fossil fuel demand—a line of policy and action.

From these visual conceptions, we eventually come to the nitty gritty of mathematics. The math within the diagram may be a simple equation, such as $y = a + x$, or thousands of "lines" of complex computer code. Math, too, in essence, is spheres and tubes. The behaviors of its ethereal "things" (parameters, variables, constants, unknowns, and knowns; the y's, a's, and x's) are correlated by their invisible links (operations, such as addition, multiplication, powers, and differentiation). The most profound relation even looks like a tube: the equal sign.

So universally applicable are these conceptual tools—the physically analogous reservoirs and flows, the more abstract control factors upon flows, and the deeper abstraction of math's parameters and operators—that clever software developers have coded them into generic modeling programs. Vivid proofs of the power of the multilevel spheres and tubes, these models for modeling can be used to simulate almost anything—ecology, electronics, earthquakes, and even retirement projections for universities. One can enter names for the things and then link elements visually on the screen in these high-tech tinkertoys. The software even stops the user with a "hey, dummy" message if the chosen math lines contradict the superimposed plan for flows and controls.

Whether designing an experimental procedure, coding a neural network, or analyzing the surface ocean's nitrogen budget, scientists can expect to reap new knowledge by skillfully applying conceptual spheres and tubes. When unknowns exist, like the magnitude and even direction (great Gaia!) of carbon flow between atmosphere and global land biota, alternate conceptions jockey for success. When the physical systems are extremely complex, like the carbon cycle, alternate conceptions can coexist—each useful for linking questions to data and theory. Scientists find themselves in many a healthy debate over values of control factors and the placement of their lines. I submit, however, that it would be as unlikely for scientists to make their models without the sphere-tube synergy as it would be for humans to gaze into the night sky for millennia without superimposing a grid of swans, dippers, and smiles.

Then too, every time we speak we display the synergy of things and relations. The dreamings, divine attributes, and science issues of the aborigines, kabbalists, and agronomists, and even the structure of many science models (there are few, if any, equations in Darwin's thick books, for example), show the capacity to generate the synergy by language, logic, and stories. The rudimentary things and relations of language, its nouns and verbs, are perhaps the neural ancestors of math's parameters and operators. Connecting words, like *and* and *or*, link nouns, verbs (now spheres in turn), and phrases; logic-toned lead-ins, such as *such as*, *therefore*, and *although*, provide flow between ideas. Larger still are our points, but sometimes the listeners or readers just do not get the connections we make. We all must draw our own conclusions.

The simplicity of such linguistic connectors can conceal subtle and profound metaphors. For example, consider a theme of this book: form and function. The metaphoric aspect here is that physically based knowledge of things and links has moved into the way we create systems of ideas. Between the abstractions of form and function, the "and" signifies not mere addition but a tube of complex ideas by which we understand that form and function are related, connected, linked.

The properties of tubes and sheets put each at the other end of a spectrum with the sphere, the tube with its linear stretch and the sheet with its planar squash. The tube and sheet can themselves form a third spectrum (grass leaves and airplane wings are both tubular and sheet-like). The three spectra can be arranged as a triangle, a kabbalah-like diagram, which, like the art of the aborigines, is both physical and mental, and which, like the carbon cycle model, is simplified yet useful. At the nodes are the pure emanations, the ideals of shape: sphere, tube, and sheet. Within the space encompassed by the lines reside all the forms of the real world and our conceptions of them.

In thinking about shape, we ask not just "Why the tube here?" but "Why the tube here, the sheet there, and the sphere over there?" These three metapatterns are best and perhaps only understood as a system. Sometimes the pure nodes are nearly expressed by life, as in the grape, its leaf, and its tendril. More often shapes are mongrels, and we split apart the three nodes with our minds. Thus a fish is simultaneously plump to metabolize, streamlined for speed, and flat to impel—a dynamic dance of functional tendencies, with the major steps drawn from the alphabet of form.

Borders in biology. Framed vertically by the bark of two species of juniper, three plant cells under the electron microscope show the intersection of the light-colored cell walls (ovoids are chloroplasts). A cap of elk hair separates the plant world from the animal world above: discarded skin of dragonfly larva after the adult has emerged, chrysalis of a painted lady butterfly, beaver dam, oak galls, wasp nest.

Borders

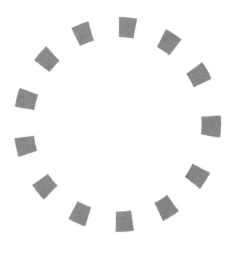

Bulwarks of Being

 At our beginning, right after one sperm passed from outside to ourside, the new zygote set about protecting itself
with spermicide. This prevented the genetic disaster of additional
penetrations of the tiny sphere's membrane—no new sperm need
apply! The membrane—a double layer of lipid molecules—is the universal sheathing of cells. Growing, we were wrapped in epidermal
cells. Still more layers swathed us after birth: blankets, bassinets, and
backpacks; warm, dry rooms; national borders.

Life at the smallest scales boasts a cornucopia of protective edges.
Origin-of-life experiments have demonstrated that lipids in water
will shrink-wrap themselves into spheres. Life has been running as
cells encased in lipids ever since. Within the eukaryotic cell the same
membrane design shrouds internal organelles—mitochondria, chloroplasts, golgi bodies, liposomes, and, of course, the nucleus. Some cells
grow additional layers around their outer membranes: silicate shells of
diatoms, cellulose cages of plant cells, chitin shields of fungus cells.
The simpler cells of the prokaryotic bacteria are also armored—with
a wall and, outside that, a capsule, both jackets of tough matrices of
molecules.

At larger scales of life, too, can be found distinct bounding surfaces. Shells encase the eggs of reptiles and birds and the bodies of sea urchins and clams. New types of borders rank among the brilliant inventions of evolution: the angiosperm seed case, the reptilian skin with scales, the bark of trees, insect cuticle. Bodies of worms and people—with pliable skins—contain organs themselves skinned with special cells, thus echoing the pattern of borders within borders of the eukaryotic cell.

In the metapattern I call borders the sheet appears in a new guise. Border sheets are the surfaces of things—of spheres, tubes, and even sheets. A leaf, for example, has a transparent coat only one cell thick, itself capped by a film of wax. A tube is often a rolled border. For spheres, the grape is again a prototype: a juicy orb wrapped by a tougher hide.

Borders function as bulwarks against forces of disruption. They cloak creatures and their internal parts against the ravages of the exterior world—the ionizing, lysing, dissolving, jolting, combusting, dispersing, bursting, rotting, eating, and crushing world. Borders hold at bay all that would destroy the difference between being and environment; they prevent universal homogenization.

Life's borders accomplish much of their bulwark functions with a simple and generic design. This design can be seen in cell membranes made of lipid molecules; in tree bark, with its tiny cellulose cages of dead cells; in mammal skins of keratinized, flattened, dead cells; also in animal hairs, scales, and feathers; in virus shells of identical protein subunits; in bird nests and beaver dams of twigs and sticks and mud. This generic design is even used for bounding the precious information contained in chromosomes, whose ends are buffered by very short sequences of DNA repeated thousands of times.

What is this generic design in biology's borders? First, to repel the drive toward entropy, a living cell or organism presents an exterior surface coated with substances already or almost dead or inert compared to the internal parts of the being. Furthermore, the coating should consist of small, repeated units. Repetition provides a sameness of surface in many directions, facilitates growth (just add more of the same), and allows bits of the bulwark to be sacrificed, sloughed away, and replaced during the maintenance of being.

We apply these rules, too, in building outside our skins the many levels of bulwarks: clothes of repeated and repairable weaves; homes with walls of plywood sheets and siding, with roofs of shingles; fences of posts and grids; and the mud layers, bricks, logs, and stones of the ancient walls around villages and cities. Cities today are bounded mainly by invisible laws, but the social institutions we live within were born and incubated safe from the entropy of continuous warfare—inside thick, lordly walls. Look at Renaissance Florence, for instance, with its craft guilds, banks, and republican government. Its mighty walls rivaled its dome and campanile in fame. Civic pride was built upon a sphere, a tube, and a border.

City walls go back still thousands of years earlier. Egyptian towns had theirs. The walls of paradise (paradise means "around" + "wall"), Jericho, Troy, and the five massive rings of Nebuchadnezzar's Babylon are the stuff of myths. The word *town* has roots in the German *Zaun*, for fence. The original Chinese word for city, *ch'eng*, also meant wall. To the Chinese empire goes the prize for sheer audacity of an ancient wall, one-twentieth of the earth's circumference.

The wall of the Kremlin. The formidability of this wall perhaps says something about the relationship of the governed and the governors.

Ins and Outs

More than two millennia ago, Ch'in Shih Huang-ti, the First Sovereign Emperor of China, dealt with big border issues. He ended the Age of the Warring States; he instituted harsh laws pursuant to the Legalist doctrine that the natural abundance of evil in people must be vigilantly blocked; he sent emissaries afar to find an elixir to wall off death; and he ordered construction of the most massive political barrier ever, the Great Wall. Many conscripted workers died in building the Great Wall, which has been called the world's longest cemetery.

Legend has it that Meng Jiang, whose conscripted husband died at the wall, journeyed there to recover his body. Discovering that he had been buried in the foundations, she wailed for the gods to help. An immense section tumbled down. Bones lay everywhere exposed. Drawing blood from her finger, Meng Jiang used its power to penetrate the bones in the rubble, and thus located the remains of her husband.

Both culture and biology express the theme of violations across bulwarks. The Khan clans punctured China's borders. The walls of Jericho and Troy were destroyed. In the Cambodian version of the *Ramayana*, the ten-faced demon Reap tricked his way across the magic boundary protecting Rama's wife Sida by posing as an aged hermit. Romeo and Juliet breached family rules as well as walls. Across the cell membrane of an intestinal villi cell sneaks a salmonella bacterium. And all beings transgress one another's borders in nature's immense restaurant.

Things are not isolated. No border is absolutely impenetrable. Otherwise how could one eat, breathe, or read? In fact, the liveliest borders are far from perfect bulwarks. They have pores not made by teeth or swords but intended as an integral part of being.

The section of wall collapsed by Meng Jiang was never repaired, and today the railway passes through. The Great Wall's most famous planned pore is the western Jade gate, or China's "mouth." It opened the way for foreign adventurers into the belly of China. From above, however, China (like all land always) is one big pore. China could attempt to shut itself off from world politics, but not from the sun.

Openings to energy sources are essential to any system that is living or contains life. Flows of energy across borders power the construction and reconstruction of the high-energy matter patterns that must be built and perpetually repaired. Some of my own research is for a NASA project called Controlled Ecological Life Support Systems (CELSS—a felicitous acronym). When built, CELSS will provide food, oxygen, and water for astronauts and, eventually, colonists on Moon and Mars. The CELSS program aims to grow crops and recycle wastes with minimal leakage of mass. Earth and Biosphere 2 in Arizona are examples of complex systems with cycles nearly closed to matter fluxes across their borders. The greater the degree of material closure in a living system, however, the greater is the need for inputs of nonmaterial energy—for example, sunlight (or electricity to make light and drive fluids)—and outputs of degraded energy, usually as heat.

For living or life-containing systems, mass fluxes are usually essential, too. Matter itself may usher in the required energy. Rejected wastes must be replaced with fresh materials. And net growth demands the addition of new mass. Across the biological kingdoms, such needs have engendered many types of exquisite pores: the ion

The Lion Gate at Mycenae. Today, pores in ancient city walls are not heavily guarded. PHOTOGRAPH BY LYN HUGHES.

channels in cell membranes, the gap junctions in the walls between plant cells, the stomates in leaf surfaces, the various apertures and orifices in animals for passing food, wastes, gametes, and matter- and energy-borne information.

In addition to their roles in life maintenance and growth, pores link parts into wholes. Within a living entity are subsystems, each with a distinctive contribution to the overall metabolism. As processors for the larger system, the subsystems exchange their inputs and outputs among themselves. Octagonal channels in a cell's nuclear membrane convey materials into the nucleus and messenger-RNA out. Body organs transfer processed fluids across entries and exits. Within the earth's technosphere, various masses, energies, and messages cross the borders of cars, ovens, radios, and computers via the inlets and outlets we have designed.

There is still another reason for pores. Stuff often needs to be stored and later retrieved. Cups, bottles, and the hollow hulks of refrigerators and trucks possess pores for transferring stowage. Cells can package and unpackage substances within lipid membranes for internal transport between parts. Storage borders vary from those of squirrel mouth pouches and animal bladders to car gas tanks. Spoons and chairs are like cups with pores spread wide.

Borders for stowage often permit the fewest possible number of pores, namely, one. Distinguishing inlets from outlets, as in the gas tank and our bladder, is yet another common design for stowage transfers. The vital flows through things that transform inflows of mass, energy, and messages into quite different outflows are often controlled and channeled across a few major orifices, too. Included here are the mammalian mouth and anus, the gullet of a paramecium, the foramen magnum of a human skull, the exhaust pipe of a car, the delivery ports of a store. When pores are few in number, they are usually relatively large.

In contrast to this pattern of a few big pores is the pattern of many small ones. We sweat through multitudes of skin pores. Ions shunt across hosts of cell membrane channels. Air holes in the shells of bird eggs, in the surfaces of leaves (stomates), and in homes (windows) riddle the borders. When the substances transferred in and out are relatively small, and the processes driving the transfers are relatively passive (for instance, diffusion), the design of many small pores can employ the border as a vast transfer sheet.

Most complex systems combine a few big holes for some substances and many small holes for others. In a leaf, gases pass through numerous stomates, light through the infinity of transparency, and liquid water with its solutes through a big pipe at the base. Our bodies mix an array of skin pores with nine to twelve major orifices.

Pores come in a great variety of themes—from few big to many small, from continually open to sometimes closed, from one-way to two-way. This exploration of borders, therefore, brings us to a level of complexity not seen in the two previous chapters, which attended to the nearly pure topological metapatterns of spheres, tubes, and sheets. Borders are systems with component patterns. They could be idealized as short tubes embedded across sheets. Functionally, they are barriers and connections, walls and bridges.

Separation and Connection

What about things that do not have skinlike surfaces for borders? How well does the pattern of borders as systems of walls and bridges hold for such things as clouds, ecosystems, and insect colonies?

Clouds lack a defining skin, yet nonetheless sometimes terminate crisply against the blue sky. Is it by edges of any sort that we recognize clouds as entities, as things distinct, worthy of notice and a name? At other times, their edges are fuzzy wisps. The complex, often shifting nature of a cloud's border, from scales large to small, is a prime example of what mathematicians call fractal geometry. Though it may be difficult if not impossible to draw the edges of clouds in an overcast sky, we posit them just the same.

In addition to the structural spectrum of visible edges from crisp to fuzzy, consider the "functional" borders of clouds. (I use the word *function* metaphorically here, as only the products of design and evolution can properly be said to have functions.) With no specialized parts as the bulwark, the functional aspect of a cloud's border is distributed among all the parts, in the overall coherence of droplets within a distinct zone of the atmosphere. A rock is similar; a split makes a new outside edge. The interior parts are just as capable of defining the edge as are the peripheral parts. Such nonspecialized borders we might call casual, in contrast to the formal borders of cells, people, and cultural artifacts.

Though far more complex than clouds and rocks, an ecosystem exhibits a barrier function nearly as casual. Not surprisingly, the science of ecology has produced waves of debate as to whether ecosystems are indeed living systems or just aggregates of living systems. In the headwaters of the river in the West where I spend the summers, bullfrogs aplenty float in the sun and bellow from shoreline haunts. Long-time residents recall the day only a few decades ago when a fancier of frog legs released a burlap sack full of bullfrogs to "seed" his pond. Their descendants soon spread up and down the river. Easily entering the larger ecosystem, the bullfrogs now have displaced the native leopard frogs.

An ecosystem lacks components that could serve as a membrane against invasion by external species. The bullfrogs merely had to wriggle into an ecological niche. Not any invading species will succeed, of course. But there is no specific functional bulwark in the thing we call an ecosystem; rather, like the omnidistributed cohesiveness of a cloud or rock, all food webs and nutrient cycles themselves serve as the barrier—a bulwark dispersed among all parts and therefore quite casual.

For energy intake, however, the ecosystem does have pores of a sort: green plants. Plants capture solar energy, which is then shunted as biomass into detritivores, herbivores, and carnivores. Other system inputs pass through pores, too; for example, nitrogen via the nitrogen-fixing bacteria in root nodules. I would call such pores formal, to emphasize their specialization, and fuzzy, to emphasize their dispersal throughout the system.

Ecosystems aside, if anything qualifies as a superorganism it is an ant hill or a bee hive. The border of each, with a few big orifices issuing forth from ground or tree, seems animal-like: barriers and pores spatially crisp and functionally formal. But when masses of workers disperse to forage, the edge of the superorganism is spatially even fuzzier than a cloud's edge. Like plants to an ecosystem, these foraging pores are functionally formal while spatially fuzzy. And here, then, lurks the dispute as to the formal status of these colonies: Do even these biological marvels merit the name superorganism?

So diverse are the ways in which entities fulfill the functions of borders, we might well think of walls and bridges as principles. The function of the bridge, as general pattern, would include the controls on the sizes and states of openness. Even more generally, bridges can include all parts of an entity that interface with the external world.

Indeed, the pore as a functional principle must expressly include "grabbers."

For instance, though light spatially enters the leaf when it penetrates the transparent outer layer of cells, it functionally enters the leaf's system of metabolic pathways only when seized by the chlorophyll molecule. Clocks first came to the geographic China when Matteo Ricci brought them as gifts in 1577, but they entered the cultural China only because internal grabbers, most prominently the emperor, accepted them as gifts.

The metapattern of border is thus a synergy of separation and connection. The skin with its protective sheath of keratin, its many small pores for sweat, and few big ones for certain senses may serve as a prototype of the synergy. But as we have seen, other mixtures of the patterns abound. Our immune system—functionally formal with its select cell types, spatially fuzzy because dispersed throughout our blood—is a vital part of the human border. The fuzzy valence shells of electrons in atoms separate one atom from another by repulsion yet allow bridges of relationship for chemical bonding. Overall, to find the functional as well as the physical borders, we can search for where and how an entity separates from and connects to others, by looking for those parts that relate both to something inside and to something external.

Border as the physical expression of both separation and connection provides a perch for further inquiry. The sheets that enshroud animals, people, and cultural containers can all be usefully viewed from

Border as interaction. The barbed spines of the seed of a weed both protect it and help it attach to the fur of passing animals.

this vantage. For example, the skins that separate animals from their surroundings can also project messages. Feathers insulate and their colors may camouflage, but, as the peacock knows, they also connect and signal. The colors of flower petals lure pollinators. Anthropologists agree that we decorated our skins long before we learned to drape them with insulating fabrics and hides. Today, few among us treat our dress as only a shelter from wind, rain, and temperature extremes. Through clothes we adjust the presentment of our souls.

In the cultural evolution of extending our biological borders, we painted more than our skins. Walls of caves and hides of huts sometimes bore pigments with meaning. All architectural surfaces both separate and communicate. Whether consciously designed or unconsciously expressed, buildings emit messages about the culture.

From ancient pottery to modern packaging, containers exhibit the binary of separation and connection. With one large opening, the pot is a basic border invention. From around the world for thousands of years, this global archetype of utilitarian enclosure has been superbly decorated.

The modern Hopis are famous for their painted pottery. About a hundred years ago the gifted artist Nampeyo (only one name) brought the Hopi pottery to wide renown. From the 400-year-old ruins of an ancestral village, she gathered potsherds, revived their sophisticated designs, and invented her own motifs, which her descendants now carry on. Through her archeology, Nampeyo became a window by which a forgotten past moved into the present. Through her creativity, she became a door for patterns to pass from the unseen to the seen, from the unknown to the known.

Doors and More as Metaphors

Visionary poet and artist William Blake penned the famous phrase about cleansing "the doors of perception." Aldous Huxley, a spherical scholar par excellence, took Blake's phrase for the title of his path-breaking book, wherein he documented the cleansing powers of psychedelics. Jim Morrison, who urged my generation to "break on through to the other side," continued the metaphor in naming his rock group The Doors.

The humble image of a door for consciousness hints at the universality of borders. The door becomes a place for passage in the world of mind. Furthermore, the metaphoric aspect of door is not limited to its pore-like quality but includes all the rich properties of borders. Indeed, linguists George Lakoff and Mark Johnson assert that borders form one of the primal "metaphors we live by." We are permeated by a linguistic edifice they call the "container metaphor."

Consider these sentences: Janice is in the house; Janice is in college; Janice is in business. This latter extends the meaning of a purely physical container, but no more so than the spatially fuzzy system of foraging bees. Now consider: Janice is in love. Thus the same physically rooted border concepts of inside and outside are also used for states of being.

Someone might ask me, "Tyler, what have you been into lately?" I reply, "Into the refrigerator, into town, and into bed." "No, I meant into, you know—*into*." "Oh, that kind of into! Well, I've really been into an obscure book called *Moksha*, an assembly of Huxley's letters, lectures, and essays about his psychedelic experiences and philosophy."

Archetypal containers. A ceremonial building of the French people, a ceremonial vessel of the Sia people, and a ceremonial shirt of the Chilkat people each separate precious inner substances from the vagaries of the environment. Each has a border that not only isolates but communicates.

One can, while reading, be inside the physical book, inside the mental book, and inside a state of enthusiasm.

When enthusiastic, one becomes literally in–god (note relation: the Greek *theos*). We embed the container metaphor in many other words, with prefixes *en*, *in*, *im*, and *em*. For example: enmesh, enshrine, encourage, endear, enforce (en + strong), environment, enchilada; incorporate, instruct (in + build), initial (in + go); impoverish; empower. Other prefixes help us *exit* from linguistic containers, with terms like escape, expend, express, discourage, discover.

Many common expressions contain metaphors rooted in physical borders, from crisp to fuzzy, formal to casual—and, above all, as barriers and pores. Examine: Deal is in the bag. Bitter pill to swallow. Hiding behind a wall of illusion. Open and shut case. In-crowd. Scattered. All ears. Loose cannon. Blurt it out. Tight-lipped. Really out of it. Maintaining face. Open-arm welcome. Airtight argument. Notion does not hold water. Food for thought. In one ear and out the other. Well connected. Imprisoned by habits. Contagious humor. Thick- or thin-skinned. Open- or close-minded. Nebulous. Flaky. Clear-headed. Tough nut to crack. Encapsulate the story. Take it with a grain of salt. Covering your ass. Impervious to fear. Living on the edge. Let's close this list soon. What do you have in mind? Thirst for knowledge. Beyond one's wildest dreams. That's it, period.

Physical systems require grabbers (or more passive allurements) to bring pieces of the outside into their theaters of action. Objects of meditation can, in turn, capture the attention and act as doors to new perceptions, ideas, and wonders. A grape can be a doorway. So can the moon. So can a doorway.

Shamans probe the doorways to alternate worlds. Through initiations they learn to pass through pores, the *axis mundi*, the world's naval. The shaman's staff, a thing, is a conduit across the barrier between worlds. Contemplatives in Chinese gardens focus on magnificently convoluted rocks with many holes, or "eyes," to remind them of the Dao. Bringing the unknown into the known, Moses and Mohammed went up mountains, physical sites, which, like the chlorophyll molecule, capture light, though of another kind.

A classic border myth is Plato's cave. Those chained to the wall inside see only their shadows cast by a great flame. One finally escapes and experiences the brilliance of true knowledge, thus cleansing the perceptions. When I visited the Paleolithic art in the caves of southern

France, it was easy for me to imagine this story's genesis. Inside and outside were literally two worlds. Rituals could amplify this contrast to create utterly different states of mind. The Sioux's rites in their steamy purification domes synchronize mental with physical drama at moments of closing and opening the door. We recycle this connection from space to mind by then imaging any shift in state of mind as a passing between outside and inside.

In many tales men cross barriers to get to their loves—Rapunzel, Sleeping Beauty, Juliet. Jesus, in a metaphor for the rebirth of consciousness, emerges from the doorway of a tomb. In an Egyptian res-

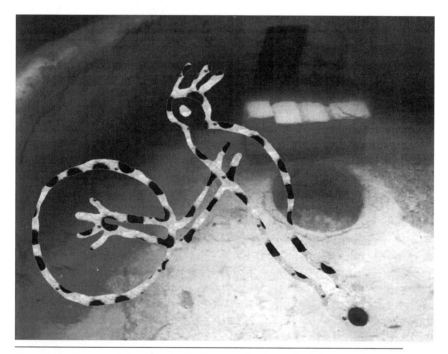

Openings for creativity. From the doorway in the wall of this round, roofed kiva flowed air that fed the fire in the large pit when the Anasazi danced and chanted. The site of transformation was through the small foreground hole in the floor—the sipapu—both ancestral and here-and-now. The Ojibwa, a shaman society in what is now Minnesota, carved into birch bark their mnemonic symbols for the order of songs in rituals; the humanoid symbol (here decorated by the author with many sipapus) sets up the chant "I am feeling for it," which is interpreted as a search for hidden medicine in the large circle (like a drum)—itself a hole in the earth.

Nampeyo. By 1935, when this photograph was taken, aging eyes forced Nampeyo to forgo the paintbrush, but her hands still shaped exquisite hollow rounds of clay with ageless primal pores.　PHOTOGRAPH BY TAD NICHOLS.

urrection myth, Isis conceives by her dead brother Osiris, whom she rescued from entombment in a pillar. In the ancient tale of Iron John, recently elucidated by Robert Bly, the king's son releases the Wild Man from his cage. In spring, Demeter's daughter Persephone bursts from the underworld into light.

Joseph Campbell's path-breaking study of the hero with a thousand faces explores the shamanic quest of crossing thresholds of understanding. Guardians block the gates in the barriers faced during one's journey. Nearby, helpers may assist the passage. Finally, the hero bursts through doors into the unknown and then brings back discoveries to be shared with all.

Such doors open life's mysteries. Through them we grow. As the potter Nampeyo has shown, art helps us find these doors. Science can too.

Between the known and unknown stand the scientists. Copernicus, before unleashing his revolutionary reasonings, thanked his predecessors "who first opened the road of inquiry." Prominent physicist John Wheeler has "made a career of racing ahead . . . and throwing open doors." And many scientists would agree with naturalist Donald Peattie that "Of our windows on the universe, science is set with the clearest pane."

Scientists investigate for information, insights, and enlightenment. To get there they need imagination, intuition, invention, and enthusiasm. Sometimes their examinations and inspections are truly inspired, and their discoveries enchant, perhaps inflame, and hopefully enrich us all.

How do scientists accomplish these border crossings? Paradoxically, by isolating parts of nature for study. Galileo framed Jupiter with a telescope to watch its moons as a system. Florentine scientists screened meat to exclude flies and proved that no maggots emerged—and thus that there is no spontaneous generation of life. Fusion physicists confine plasmas in magnetic bottles. Physical containers are often vital: test tubes, flasks, and beakers; nets for ocean plankton and pipes for coring into ocean sediments. Experiments delimit external influences; the scientists control the variables. But experimenters also create refined bridges across such framings: the tubes of manipulation inward and observation outward.

I still recall the flash of delight I felt as a student during a lecture when I was struggling to learn the theoretical engineering tool called control volume analysis. It was not the arcane intricacies that moved me; rather, the insight that the technique consisted of little more than slapping imaginary conceptual borders around a system. Control vol-

ume analysis thus makes explicit inside and outside zones, and it spec-
ifies a border across which all fluxes in and out must be tracked. There
is no absolutely correct place for drawing these analytical lines, only
places of more or less utility in the subsequent calculations.

Modelers such as myself may lump all marine waters into a single
bounded "box"—or three, five, seventeen, or even thousands—and
then compute the crossing fluxes of these control volumes. Climate
modelers cover the mathematical Earth with grids of such boxes. In
the workings of science, barriers and pores, walls and bridges, are
probably isomorphic to the synergy of spheres and tubes.

Thus borders appear to be fundamental in language, in myth, and
in science. They extend into every sphere of life. Borders greet us
today as the very substance of our most difficult and troubling problems, solutions to which will require passage through the doors from
known to unknown.

Border Mind

Respectful visitors to Liu Ling, one of the Seven Sages
of the Bamboo Grove 1,700 years ago, were affronted
by him sitting naked in his room. Their discomfort was stoked further
when he would announce, "I take the whole universe as my house and
my own room as my clothing. Why then do you enter here into my
trousers?"

To reach such a Daoist unity with all existing things, we play,
metaphorically, with our selves and borders. And well we should. We
invent them as easily as fish swim, from highway lane dividers to
Jefferson's call for a wall between religion and government and
Madison's idea of freedom of speech as a "bulwark of general freedom."

Perhaps in certain lines of chalk, paint, and tape lie the purest
expressions of our fondness for edges. Outside watch enraptured spectators. Inside, bound by the rules of the field, the players finesse balls
through pores; some small and crisp, protected by threshold guardians
like goalies and sand traps; others large and fuzzy, the holes in the
opposition's strategy.

Materially insubstantial but full of weighty laws and protected by
internally dispersed cadres of warriors, political lines are not so different. Unfortunately, spectators in the vast historical past have too often

found themselves suddenly under new rules inside their neighbor's expanded field and forced to push back on the bloody line of scrimmage.

To appreciate the sundry zones circumscribed by disparate levels of legislation, look at maps. What shapes are found? Circles? Rarely. So an absolutely minimal perimeter must not be determining. Rectangles are stylish, though. Simple to survey and easy to pack in groups or to

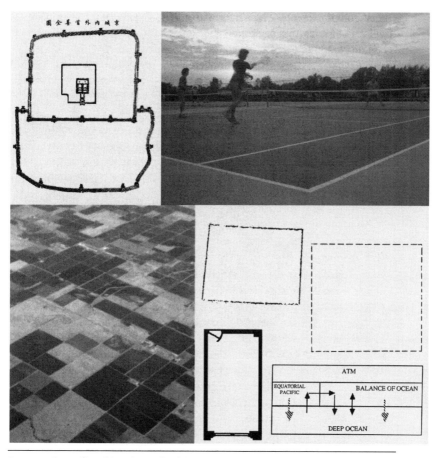

Rectangles of cultivation, fortification, legislation, computation, and illustration. *Clockwise:* agricultural fields; Beijing's nested series of ancient walls; modern ball court; the county of Stonewall in Texas; a three-box ocean model (with atmosphere) for carbon cycle studies; author's office walls. Remaining is the State of Colorado, encased by the conceptual fences of latitude and longitude.

subdivide, rectangles range from gardens and suburban plots, to city blocks, counties, and even states. The ancient Chinese said *tien yuan, ti fang*—heaven round, earth square. Wherever the land permitted, they built their walled cities in rectangles.

Constraints in the lay of the land can force odd angles and convolutions in domain boundaries. In the eastern United States, state edges often abut rivers. With more ecological sense, the federal wilderness boundaries tend to follow the upland edges of watersheds. Truly weird shapes can result from political pressures; for example, gerrymandered election districts. Once, while driving near the outskirts of Houston, alternating signs flashed by me without visible change in the landscape: Entering Houston, Leaving Houston, Entering Houston, and on and on.

Should we rank the concept of legal lines—with their diverse controls for what goes in and what goes out—as a major border invention, like the cell membrane, bark, and skin? Undoubtedly. Legal lines, however, are mere fledglings compared to the borders of biology. With barely tested wings, they present many deep problems.

Our legal limits do not just enclose us, they are us. What happens when a higher-level government makes a ruling that conflicts with local interests? Vexing issues and lasting disputes arise from the multiple bounded arenas we live within and identify with. Fuzzy-edged cultural and business groups, each with a distinct agenda, permeate all levels of the comparatively crisp land-based lines. Laws apply property concepts to our bodies and our extended skins of homes and habits in a plethora of privacy issues. At big scales, nations accost one another by entering each other's trousers, by physical force, unfair trade, and nearly unstoppable cultural incursions.

An even larger issue concerns the borders between society and nature. Economic exigencies pound on the undefended, casual borders of ecosystems. Wild, inspirational nature is shrinking fast. Where nature was once the big outside to us, we now wrap organisms within the jurisdictions of nature "preserves" and "refuges"; thereby, to a large extent, determining their patterns of numbers, types, and even their ability to evolve. The ranges of some species have been reduced to walled habitats in zoos. What remains of pristine nature we surround and protect in small enclaves, pockets of wilderness. William Blake warned against such confinement:

> Robin redbreast in a cage
> Puts all Heaven in a rage.

Cultural cages around nature can be quite fuzzy. Our trash diffuses into nature as countless cultural tentacles. I watch it invade the undefended wilderness as easily as exotic bullfrogs:

> Beer can by side of road,
> Gets me ready to explode.

The tiniest part can serve as doorway to the whole: As I sit on the porch, a nether zone between indoors and out, my thoughts probe the remembered beer can as a pore to an immense space filling with all our unmanaged excretions with no happy place to go: excesses of carbon dioxide, urban ozone, oxides of sulfur and nitrogen, noxious chemical wastes, undertreated sewage, nuclear reactor wastes; cities bursting reasonable confines in cancerous, fuzzy, ever-growing sprawls, making for further and further commutes that burden the planet with more pollution; the ever-glaring, ever-blaring lights and sounds as a thick gel that walls us off from the stars and from silence. As I seethe, contemplating these border issues, a fly persists in landing on my arm. A slap, and it falls dead to the floor. Guilty at the sight, I kick it off the edge. The human border truth: out of sight, out of mind. Well, Confucius declared that *fang-wai* (outside the square) was outside his concern.

Border issues do not usually lend themselves to easy resolution. By what rationale do I justify killing a fly but try to limit my meat consumption to reduce the number of gentle brutes we raise to slaughter? Actually, that one is easy: my beef against beef is primarily in protest of the stomach-turning overgrazing I have witnessed on our public lands. And, hey, that fly probably owes its existence to a cow pie. The group Ducks Unlimited has been a major conserving force for wetlands, so that its members can continue to violate the skins of millions of ducks each year. Moral issues are usually border issues. And we must look inside ourselves, to our deepest values, and examine the many levels of lines—fuzzy or crisp—that we draw.

Designers of China's fabulous walled gardens contrast the crisp, blank, often white walls with the intricate, indefinable shapes of plants, water, and rocks thus enclosed. The result: perfect landscapes for contemplating the unbounded, ineffable, ambiguous, and powerful Dao. This ability to see boundlessness despite walls comes from the insights of mountain-dwelling poets, philosophers, and visionaries, who looked inside by going far outside.

> You ask me why I dwell in the green mountain;
> I smile and make no reply for my heart is free of care.
> As the peach blossom which flows downstream and is gone
> into the unknown,
> I have a world apart.

Li Po pulled away from the binds of society in order to connect with peach blossoms. Mystical borderlessness is one possible gain. The greater the variety of border patterns we contemplate from nature—the generative matrix of borders, formal to casual, crisp to fuzzy—the greater are the possibilities for expanding our minds. Sensory border patterns provide templates for those mental, as in another couplet by William Blake:

> The wild deer, wandering here and there,
> Keeps the Human soul from care.

Freedom from care, however, must not imply lack of care. Adjusting where and when not to care is an ongoing border issue of life. Too much borderlessness, as parents discover, breeds unhappy children. The young grow best with free, unbridled play within known and consistent behavioral bounds. The self erects a shell during development, which, like that of an egg, includes both barriers and pores—barriers against coercive onslaughts and overidentifications; pores for learning, expression, and love.

On the collective level, we incubate our society through the generations within cultural shells. In what ways do these fuzzy matrices of technology, language, art, science, and architecture, and of values, laws, and myths, separate us too harshly from nature? In what ways do they connect us to nature via bridges unachievable without such sheltering from the storms of time's brutal uncertainties? The inquiries out into nature we have launched from such shells reveal new borders that serve as paradigms for the next stages of cultural evolution.

Today, analyses of the cycles of essential biological elements, such as carbon, oxygen, sulfur, and nitrogen, have shown that ecosystems, with their materially wide-open borders, are themselves subsumed into larger and larger systems that encompass the cycles of atoms, that effect closure, and that thereby take the pattern of the border to Earth itself. At the largest scale, the biosphere has skins.

Beneath our feet, the sheet of soil, life's creation, sequesters nutrients and water and mitigates exchange between the underlying rocks and life. The soil is a skin of the biosphere. And look above. In Jim Lovelock's words:

> The atmosphere is not merely a biological product, but more probably a biological construction; not living, but like a cat's fur, a bird's feathers, or the paper of a wasp's nest, an extension of a living system designed to maintain a chosen environment.

The sheets of soil and air do seem to be actual membranes of the largest living system, the biosphere. These two sheets do not merely exist alongside the biosphere; they were created by and are now intrinsic parts of the biosphere.

Layer upon layer we have extended our selves with cultural shelters: clothing, buildings, cities, and nations. Now we can identify with Earth's outermost membrane as shelter, too. Connecting all our invented squares is one great and sustaining sphere of air. As Liu Ling might remark today, I take the whole Earth as my body. Why then are my skins being defiled?

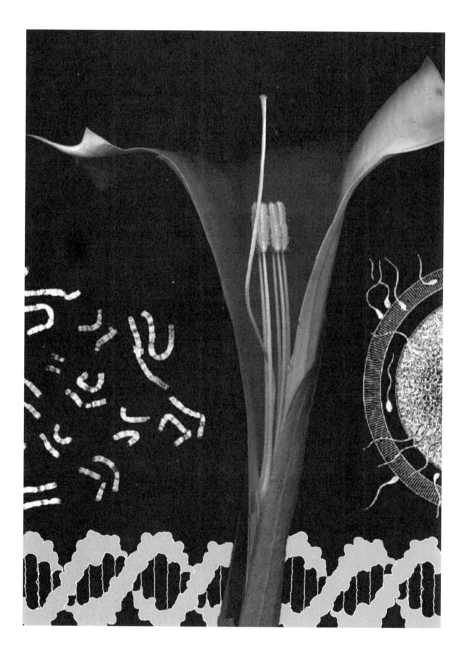

Binaries in biology. Nature has selected for binaries at all levels of reproduction: stamens and pistil in the flower of Sacred Datura, a mammalian egg cell and suitors, pairs of human chromosomes, and the double helix of DNA.

Binaries

Triumph of Twos

 Last night I was watching aquatic ecology in a small hollow along an otherwise dry arroyo. Water beetles swam in a three-dimensional world, propelled by two long legs as oars. Suddenly a beetle with four oars, two per side, glided into view. I peered through the glassy surface more intently and discovered a couple, mating. The male, above, did all the paddling, and so their unity swam like the more numerous loners. Binary is surely the sexiest metapattern.

The ancient Daoists took male and female as one basis for their universal philosophy of the yin and yang. This most famous binary symbol of all—a two-toned, Escher-like embrace of tadpoles—was, according to one suggestion, derived from clouds, with their mutual sides of shade and light. The sky reveals another yin & yang tossed out by the cosmos: the identically shaped and sized Moon and Sun. And the early Daoists would certainly have clapped in affirmation, had they been able to gaze into microscopes at sperm and egg. The passive, yin egg is rolled along by fallopian cilia; the active, yang sperm swims. With higher magnification one can witness even smaller genetic objects that come in pairs, the chromosomes. And still deeper, electron microscopes reveal that each unraveled chromosome contains a huge molecule structured as a double helix, biology's insignia. Its chemical rungs contain the genetic code, whose units are a pair of pairs (adenine with thymine, cytosine with guanine).

Joining the Daoists in cheers at these infinitesimal twos would be specialists from another ancient tradition. To the alchemists, moon & sun and female & male were examples of the stupendous principles of sulfur & mercury, which formed the basis for a theory about the generation of metals. Had alchemists access to the current findings of their rebellious descendants in the lineages of knowledge, they would have winked in recognition. Chemists and physicists, wherever into nature they dig today, uncover pairs by the pailful. A prominent chemist recently declared that essentially all chemical reactions occur between a base and an acid. A physicist's diagram of light's electric and magnetic fields looks somewhat like the alchemical insignia of intertwined snakes. Positive and negative charges stabilize atoms. Two types of massive particles—protons and neutrons—form the atomic nucleus.

Unlike Daoism and alchemy, however, science posits no overarching generative pair. Its closest foray into such temperament comes

Binaries in Daoism, alchemy, and physics. Yin & yang and moon & sun here serve as the poles that generate an electric or magnetic field.

from the modern physicists who spend lifetimes pondering the subtle twists of twos in that realm where math and matter are indistinguishable. One, Niels Bohr, had the yin-yang embrace carved on his gravestone, to honor what he called the principle of complementarity. He believed in a deep basis for such pairs as subject & object, the wave & particle nature of matter and light, and the uncertainty principle's alternate pairs of position & momentum or energy & time.

Biologists, too, have made use of the binary as ideal. In the heated quest to unravel DNA's structure, James Watson decided that he and Francis Crick should focus not on triple but on double helixes because "important biological objects come in pairs." Watson's personal rule of thumb is no match, however, for the truly universal concepts believed by Daoists and alchemists. Modern science, with its divorced disciplines, almost by definition has nothing to say about the big All. It has dismissed the yin-yang hypothesis and the mercury-sulfur hypothesis. The pairs of science—protons & neutrons, adenine & thymine, and so forth—are each unique and reside in unique contexts within the spheres and tubes of things and relationships. The complementarity principle of physics is not an explanation for sex. No political scientist would invoke the necessity for the two-party system by saying it simply follows on the universe's penchant for pairs.

For good reason modern science contains no overarching principle of binary. First, we in science must pluck the fruits of knowledge by taking things as they are, not fitting them to projected procrustean concepts. The Daoists may have cast their own heads in the clouds for the yin and yang. The alchemists may have been high on yellow and red vapors when dancing with the two sides of Hermes. Yes, we scientists have abandoned the pretense of attempting to explain the big All, but what we have gained in powerful and practical understanding of a plethora of details is in my view well worth the trade.

Twoness, therefore, is not an idea by which we can deduce what lies behind the curtain of the unknown. It is not the theoretical beginning to a course of study. To counter such dictums as the alchemical slogan of "as above, so below," the now venerable organization dedicated to science, the Royal Society of London, made its motto "no faith in words."

And yet—despite all denials of a universal twoness, a bounty of twos is now apparent across the fields of existence. Consider: the proponents of "no faith in words" discovered "every action has its equal and opposite reaction." And as Carl Jung showed, the binaries of alchemy and Daoism are ultimately insights not into matter, but into mind. From the grand symmetries and constructions of physics, to genetic couplings and the two-party lock on politics, the ubiquity of binary at least hints at a potential reservoir of properties. In probing why twoness is so far-flung, this metapattern becomes a framework for asking fruitful questions.

One question is, To what extent are there types of binaries? For denying any universal binary does not require, in rebound, declaring that each binary is absolutely unique. Is there a middle ground beyond this logical binary of all or nothing, a spectrum between the extremes wherein we might find classes, types, families of the metapattern of binary?

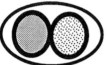 ## Families of Simplest Complexities

We have already seen that the metapattern of sphere comes in families—subpatterns of the main. For example, some spheres exist passively as a response to omnidirectional forces; others are formed for a function—to minimize surface area or to maximize strength. The tube also has families, such as those that minimize end exposure and those that transport matter. Types of borders include the crisp and the fuzzy, the formal and the casual. Any particular sphere, tube, or border, of course, may belong (and usually does) to more than one family.

These subpatterns of spheres, tubes, and borders give rise directly to several prominent families of binaries. One such offspring in the genealogy of shape descends from the mother of all patterns. A sphere (and not just the idealized shape, but as symbol for any entity) cuts existence into two parts: inside and outside, thing and the rest of the universe. The two—entity & environment—relate, determine, create, and influence each other in a radially linked dance by which any thing is joined to its surrounding sea: ship & ocean, particle & crystal, galaxy & cosmos, genome & cell, organism & ecosystem, computer & network, person & society. The figure & ground binary forms a basis for logic and language, in classifying entities and giving them context, and as a basis for science, in investigating the activities of things in their environments.

We have woven this simplest complexity of entity & environment into many of our most subtle dualisms, into conceptual systems, into debates of meaning, into tools for describing. Combining a shake from the inside with a shake from the outside has created a great many dishes across the disciplines. These include (listing first what I consider the "inside" example in each pair): evolution's variation & natural selection and genotype & phenotype; psychology's nature & nurture, unconscious & conscious, and introvert & extrovert; gestalt theory's figure & ground; sociology's in-group & out-group; math's differential equations & boundary conditions; philosophy's subjective & objective, rational & empirical, and reductionistic & holistic. In entity & environment are grounded such ontological binaries as in & out, near & far, here & there, and such activity binaries as action & reaction, stimulus & response, toward & away, together & apart. The Beatles' haunting song "Within you and without you" explores a personal view born from this binary.

The sphere is not the only metapattern whose presence begets a family of binaries. So does the tube. By itself, the tube offers a number of potentials for simple binaries. Its two ends may be differentiated and thus become partners in coupling, as in the so-called male and female ends of hoses and cables, and the axons and dendrites of neurons. The ends of tubes often form a system by directing the flows of matter, energies, forces, and capabilities, such as the headwaters & mouth of a river, the mouth & anus of many animals, the capital

Four major types of binary. *From left to right:* sphere & ground; two ends of tubes; the sphere & tube synergy; and two spheres—perhaps the prototype of all.

& pediment of a Greek column, the shoulder & elbow articulations of the humerus, and the creation & destruction tips of a pencil.

In addition to these families drawn from the ends of tubes, if the two possible directions of flow are isolated, the very lengths of tubes lay down another type of binary. This family includes many famous doublets of flow: veins & arteries, xylem & phloem, efferent & afferent nerves, divided highways, question & answer, addition & subtraction. This last can be visualized as movement back and forth along the number line. Might gradations, therefore, such as big to small, wide to narrow, hard to soft, and kind to cruel, be examples sprung from a primordial tube binary in the mind, serving as an archetype for such quantitative and qualitative scales?

Although spheres and tubes individually act as templates for a variety of binaries, when placed together into a system they form one of the most powerful binary families. This family, from which shamans and scientists build diagrams, I earlier called the sphere–tube synergy. It is the binary of things & relations.

A sphere & tube marriage of metapatterns, for example, permeates language: birds fly, rain falls, love awakens, Earth spins. Modifying these nouns & verbs are adjectives & adverbs, and all are coalesced into what linguists call noun and verb phrases. Things & relations are repeated at a smaller scale within words, as consonants & vowels. For instance, deaf babies "babble" with their hands in two ways: static hand shapes and mobile gestures, which linguists believe are the rough equivalents of consonants and vowels.

So we speak with nested levels of the pattern of things & relations. No wonder, then, that Descartes claimed he could explain the universe with the binary of matter & motion. Newton conceptualized masses & forces, Coulomb experimented with charges & fields. Modern physicists now speak of real & virtual particles, the things and forces among them. The specifics change, the binary remains.

Perhaps the sphere & tube binary can help explain what impels us to think of form & function as such an intimate pair. Form is the sphere, the thing; function is the tube, form's relations. To know form requires freezing a thing in space; for its function one needs information about its activity, in other words, its pattern in time. Thus a bridge of common conceptual ancestry may link the pairs of space & time and form & function.

What more? Do space & time, form & function, consonant & vowel, and noun & verb have other siblings in the sphere & tube family? The congruence may not be exact, of course, but other pairs do suggest a common theme. In physics: matter & energy, particle & wave. In biology: plant & animal, egg & sperm. With this family of binaries, we may be getting as near as we can hope to an underlying twoness, an updated yin & yang.

A candidate for the most prototypic binary family is a particular case of the sphere-tube synergy. Connect two spheres with a tube and focus on the spheres as a binary (this is an elementary way to visualize a binary). Physically, two equal spheres bridged by a tube can be balanced upon a central point. Conceptually, we model reality as many levels of balance. And balance is inherently binary, whether in a balance of power, a balance of nature, a balance of opinion. Daoists aim to balance their yin & yang, Jungians their anima & animus.

A sense of equality between two sides, in a harmonious nexus, pervades many of our most common binaries. For example, look at salt & pepper, Burns & Allen, inhale & exhale, sun & moon, man & woman. But despite the necessary degree of symmetry in these binaries of two spheres, an element of asymmetry is just as vital. Consider a pile of apples balanced by a one kilogram weight. The utility of this system comes about because of the equality created between unlikes; the weight can be used again and again with any number of products.

Vice President Al Gore equalizes unlikes in his book, *Earth in the Balance*: "The key is balance—balance between contemplation and action, individual concerns and commitment to the community, love for the natural world and love for our wondrous civilization." To me, the power of such ideas flows from transforming—through a balance metaphor—the obvious asymmetry of each binary of figure & ground into the more symmetrical binary of two spheres. Choosing contemplation & the individual as the entities, then action & community are their respective environments. (Take your pick on how to map nature & civilization onto entity & environment.) The macro and micro are both made into spheres placed on a seesaw, and the life that goes on within you and without you is in balance. Attributes that embrace the polarities of symmetry & asymmetry, equality & inequality, sameness & difference achieve a Zen-like and ever-dynamic balance.

Striving for balance in opposites or polarities means that one runs the risk that the two can slip into conflict. Negotiations fail; the issue goes to court, where justice not balance is the goal, or the verdict is worked out on a battlefield, winner take all.

Two spheres engaged in competition form a subtype within the family of two interacting spheres. Witness the momentous clashes of history in the epic-sung battles of the *Iliad* for the West and the *Mahabharata* for the East. The titans have since fought again and again: Athens & Sparta, Guelfs & Guibellines of medieval Italy, the U.S. internal war of North & South, the Axis & Allies of the Second World War. Jean Houston claims the root archetype here is the battle of two brothers, the Hebraic Cain & Abel, the Egyptian Seth & Osiris, the African Nommo & Ogo. The generations, the sexes, polarized social groups such as rich & poor and workers & management, and products like Pepsi & Coke can all get into or be made into archetypal rivalries.

The medieval rivalry between Florence and Sienna ended with Sienna virtually wiped off the political map. But in modern football competitions, such as the Superbowl (or as my six-year-old niece termed it, the Stupidbowl), the clash of titans ends with both teams "winning"—laughing all the way to the bank—and with a chance to face one another the following year on a level playing field. When competition is institutionalized as an ongoing binary, opponents compete within a more encompassing level of cooperation.

The evolutionary biologist Richard Dawkins has forcefully described the progressing cycles of advanced designs that emerge from the "arms races" in evolution, such as the binary tournaments of predators and prey. A final harmony may occur, akin to the merging of the red Genji standard and the white of the Heike sect into the Japanese flag. But sometimes continual vying is itself the desired end. To do away with Superbowl Sunday in America would be as (or more) unthinkable as putting an end to Independence Day celebrations.

Do we walk through woods and see nature as red in tooth and claw or as plants giving oxygen to animals, fungi, and bacteria and then getting carbon dioxide in return? Do we experience city streets as battlegrounds of cars and bicycles and pedestrians, each trying to get ahead, cut across, or do we see them as busy corridors where vast networks of people and institutions accommodate one another—even more, cooperate just to get the food to market each day, to haul away the trash, to keep the whole crazy thing going? Why do we recognize the concepts of cooperation and competition anyway? Do they form a fundamental binary in our minds? If so, do they compete or cooperate for attention?

The concept of dynamic balance, naturally, is open to multiple interpretations. I do not believe that the functional properties of binaries, in general, will ever be as crisp as, say, the transport functions of

Binary as conflict. An archetype of binary as conflict is man overthrowing man, as exhibited by Herakles and Cacus in the Piazza della Signoria of Florence. Such battles echo the intense competition between males of the same species throughout the animal kingdom. Young male sea lions tiff, practicing for future rivalry. Even more general is the binary of predator & prey. And as spiders entrap their flying prey, so we speak of opponents in chess falling into snares.

tubes. Many binaries are, after all, in the inner eye of the beholder. One may perhaps favor a binary between Madonna and Arcturus, but to most of us these two have little in common besides being big, hot stars. On the other hand, many binaries, such as the sexes, do seem to possess an objective side. So if it walks like a binary, and talks like a binary, then . . .

Another concept applicable to binary is synergy. Thinkers ranging from the anthropologist Ruth Benedict to the engineer Buckminster Fuller have found the concept of synergy a powerful tool for understanding their respective worlds, and synergy is behind much of the current and well-justified hullabaloo about emergence of complex properties from simple interactions among simple parts. While dynamic balance evinces the parts, synergy points directly to the whole created by the parts. Synergy offers an additional perspective.

Buckminster Fuller touted the word synergy and took the yin-yang as its symbol. One of his main examples of synergy was metal alloys, which can be stronger, harder, or more resistant to oxidation than any of the constituents—or even the sum of the attributes of the parts. Fuller formally defined synergy as "the behavior of the whole unpredicted by the behavior of the parts taken individually." Today a rallying cry for the proponents of holism is that "the whole is greater than the sum of its parts."

But the draw of holism versus reductionism as the chief god of discovery need not degenerate into an us & them conflict. The strengths of alloys can be at least partially predicted with the tools of quantum chemistry. Very little of importance is ever just the sum of its parts, except money. To me, reductionism & holism are two sides of a figure & ground binary that reside in the act of distinguishing any system.

Whatever its problems, synergy is still a magnificently evocative word. When synergy slips from the nooses of dogmatic definitions, devotees, and detractors, it calls attention to the process of parts forming wholes in all its diversity. Synergy is what the toddler witnesses, awestruck, when she first places that second block to make a column, and then a third. Synergy is undoubtedly at work in the complicated evolutionary forces that gave rise to sex and have stamped it with the seal of approval.

And what is the minimum necessary to yield synergy?

Right.

Of the synergy of binary, I have several favorite examples: A humble faucet mixes hot and cold waters, each a single thermal point, but voila!, a spectrum of temperatures. Architects have combined concrete's talent in compression with steel's ability to withstand tension; reinforced concrete now shapes the urban landscape. A prototypic synergy in biology is the rock-hugging lichen, anchored and protected by its fungal matrix, capturing the sun with its algal half—the

Binary as union. An archetype of binary as union or synergy is woman & man, here in a sculpture in Florence. Moths practice their own kind of eros. Lichen, too, is a coupling—an interspecies binary combining algal and fungal parts into an ecological system with a leafy structure all its own. Union is evident in this early water faucet for mingling hot and cold flows.

textbook case of symbiosis. In mythology, Ariadne added her thread to the sword of Theseus, and thus together they defeated the Minotaur.

If dynamic balance is the relationship of the ingredients of binaries that express synergy, then what is the driving force behind the pattern, the why? I believe it is helpful to introduce another concept here. It is parsimony. Parsimony makes a pattern not a oneness or threeness or fourness, but a twoness.

Binary is the minimal system, the simplest complexity. There is no newness without at least twoness. Binary can not only extend a thing in space but transcend it in type. One plus one can equal a new one, bigger in mass and enriched in attributes. Because two is the minimal number to get something new, binary is a design attractor, a pattern into which systems tend to form and stabilize over time.

The parsimony explanation can apply wherever the weeding of a diverse field of shapes creates systems over time with select properties. In biological, cultural, and technological evolution, the sequential process of invent, search, and select can yield binaries. Take sex, for example. New theoretical work has shown that, above one, two sexes is the most evolutionarily stable pattern. Twoness of gametes obtains the fitness advantages of genetic recombination, without the increasing penalties of genetic dilution that would be the outcome if even more sexes were involved in producing a single offspring. The fact that two provides a dynamic interaction without overbearing complicatedness may also be why many of the most stable democracies show trends toward two major political parties.

When any evolutionary process makes patterns, at least a modicum of efficiency is achieved. Binary is the most economical means to get complex wholes with nontrivial new properties. The crucial step is taken in going from one to two. In contrast, progressing to three or more could be improbable, redundant, or a hindrance to stability. Three water lines fed to a faucet would provide little advantage over two. A third political party would add just another screaming voice denouncing the current policy—and worse, might skew an election to the least favored of the two main powers.

To be sure, increasing a system's components from two to three may have benefits. I suppose we'd have slightly better depth perception with three eyes and more choices on life's hike with three roads diverging in a wood. Nevertheless, the quantum leap comes via the difference between one and two. With only one eye, no depth perception; with only zeros, no code; with only one road, no choice.

Given all these structural goodies owing to binaries, it would not be surprising to find binaries haunting the very roots of thinking. Binaries help us. We love them and nurture them because they are the simplest way to get complex, and therefore the original way to make thoughts.

Recently my dentist affirmed one of my obvious self-diagnoses with the quip, "If it walks like a duck, and talks like a duck, then . . ." (I yawned). But the phrase started to repeat in my head in a lot of situations without ducks, and I couldn't figure out why a tale so silly had become an obsession. Then it hit me—Gregory Bateson's "double description." It's the jump from one observation to two, and thus to profundity. Walking and talking like a duck gives simultaneous knowledge to two senses. Double description is essentially any experience of multiple converging abstractions. Bateson and his anthropologist daughter, Mary Catherine, have posited double description as one of the founts of mind.

Remember when you were young, working on your sums, adding a long column of multidigit numbers by hand? Some errors can be caught by repeating the addition with the standard procedure of top to bottom. But that was likely to iterate errors as you fell into the same mental ruts twice. The really foolproof check was to reverse the procedure, summing the entire set by starting at the bottom numbers and moving up. That's double description.

I've been applying double description in my mathematical modeling. From fine to coarse levels, used dozens of times a day, double description is a staple in the computational toolbox, the only way to wend one's way into new digital landscapes. Have the computer program ready to calculate a thousand or million times over phase space. But first print only one answer. Check it by hand computations. Do they agree? Then predict the direction of, say, net photosynthesis—only the change, up or down—as carbon use efficiency goes down. Now does the next computer run corroborate the prediction? Compare and create a dynamically balanced synergy with parsimony between computer & hand, then computer & mind. Then sit back and watch the screen begin to produce things you couldn't have predicted—but that you can now observe with confidence.

Across Parallel Pairs

The preceding tour of families of binaries took in spheres, tubes, spheres & tubes, and coupled spheres. I have held back another metapattern, borders, until now because its binary family in the genealogy of shape provides one of the clearest routes into the topic of parallels across binaries. The focus shifts from the connections between halves of a binary to those between two or more binaries.

Once in the mesa country of Arizona, at the edge of a wide ravine, I saw through a window at Paolo Soleri's visionary community of Arcosanti the parallel between the inherent binary of a border and the concept of figure & ground. This particular window, like a Chinese garden's moon gate, framed a circular landscape. Unity was dual: earth below, sky above. At the line between distant cliff and sky, spindly trees stretched into the air. The trees—binary beings at the border of earth and sky—probed with fractal fingers into both realms, pumping ions from below and photons from above, reaching down with one half into darkness and moderated conditions, up with the other half into contrasting and often sharp extremes.

Over the years another shape—a garden hoe—has, for me, joined the tree to pose a koan: how are hoes like trees? Well, with specialized earth halves they both penetrate the soil. Yet another solution spins the hoe a half rotation, making its handle more akin to the tree's roots: the hoe radiates from our bodies, held by hands, as trees from the earth held by soil.

Terrestrial plants are a skin between earth and sky. Culture is a skin between humanity and nature. Its tools and machines (pen, spoon, tractor, airplane) are binary because they stretch between our bodies and the environment.

Are all borders inherently binary? Our skin has an epidermis (air-side) and dermis (body-side). And membranes of cells are double layers of lipids. We seal buildings and cars with primer for the entity-side and paint for the environment-side. The structural and functional match that borders must make to both inside and outside drives them to be binary.

I will call such matched binaries *parallel pairs*. Envision each binary as a single rung (a tube binary) on a ladder, arranged so that the ladder's parallel vertical poles that bridge the ends of the many rungs are the matches across the corresponding halves of the two binaries. I can think of many parallel pairs; the ladder stretches across a vast distance, for the pattern of parallel pairs is widespread and goes far beyond the example of borders.

In general, things will show parallels when their creation brings them into contact with preexisting binaries. For instance, day & night is imprinted on our retinas as two types of photoreceptors, the cones & rods. Like plants, many animal bodies are polarized into earth & sky halves, with soft, light-colored bellies and hard, dark-colored backs, primarily mapping the relative absence & presence of predators from below & above. The motion of animals polarizes their surroundings into toward & away, an effect that is then simultaneously a cause for

The earth & sky archetype. The temptation to create a parallel between the earth & sky and our two hands led to the stances of the remarkably precocious Buddha at his birth and Christ on the day of judgment. Plants are the original masters of the two worlds—here represented as the view out a circular window at Paolo Soleri's Arcosanti.

their body's evolutionary shaping into sensory front and trailing rear. Given up & down and front & rear, right & left follows for free.

As well as in the surrounding environment, the preexisting binaries may also reside in the very small. Positive and negative charges establish, in the main, the dynamics of other numerous pairs in chemistry's seething cauldrons: ionic salts, polar molecules, anion & cation, acid & base, oxidation & reduction, anode & cathode. Since other atomic patterns, such as the up & down of electron spins, also contribute to chemistry, not each parallel is a simple ideal like the ladder image. Permeating all, however, is positive & negative.

Another kind of parallelism comes from considering how the magnetic domains of stored computer data connect to the binary symbols of codes. Unlike the relational poles across the eye's cones & rods to day & night, the 0 & 1 of the binary code of software could be strung, in theory, to either the polarized or unpolarized magnetic domains of hardware. The parallel must be made; how is arbitrary.

From parallels simple and rigid to arbitrary and swiveling one can explore a vast landscape of wholly unrelated binaries. Earth & sky is not particularly related to up & down electron spins (except in our language). The DNA double helix and the pairing of chromosomes seem to be independent inventions, each created by the parsimony of two in their own functional regimes, first for the duplication and then for the recombination of genes.

A binary that is independently invented might seem to have little occasion for parallels with other binaries. Yet because of the need to harmonize with many external and internal factors—specifically, other binaries that preexist—inventions usually exhibit some degree of parallelism, along a spectrum from strong to weak. Within this spectrum, the strong parallels are the most intriguing, beckoning us to follow them, as dogs sniffing along fresh tracks.

In many sexual organisms, the male gametes are tiny, motile, numerous, and tubular; the female ones are big, stationary, few, and spherical. Was such a split driven by some inherent functional synergy in shape's spectrum? We find the same binary of motile & fixed in the ranging behavior of the adults. In many animals—such as lions, wolves, hyenas, and macaques—the males must leave their birth groups and seek entry into other groups to mate. "Lord, I was born a ramblin' man"—like sperm, like pollen on the wind.

Settled at the home dens, females nourish the young, the spatial stability of their eggs writ large. Of course exceptions exist. For example, male birds often secure a nest site as lure for the females, and chicks are fed by the roaming forages of both males and females. Nevertheless, the binary of motion & form, or tube & sphere, has imprinted a noticeable degree of parallelism from the tiny gametes to many adult behaviors, including ours.

Close to the dome homes, our female ancestors tended the children and the local plants. Away over the landscape, the men sought protein and hides. Society derived from this pairing two types of experiences: distant & local, spear & basket, hurling & hoarding, hunting & gathering, killing & birthing. The tendencies of modern men and women, evolved from ancestral behaviors, have been reinforced and further evolved through technology and language that parallels the ecological binary of the restless & rooted, the animals & plants.

Psychologist Carol Gilligan has discovered a binary that looms in the moral development of girls and boys, what she calls the moralities of care and justice. And linguist Deborah Tannen has observed a broad binary in the conversational styles of the two sexes: the lecturing manners of men contrast with the individuated, questioning, and back-and-forth styles of women. Tannen distinguishes the two as report talk versus rapport talk.

Perhaps the dimorphism of hunting & holding helped form these binaries in moral codes and conversational styles. The dangers of killing mammoths and the need for all participants to follow the game plan may have given rise to life-and-death laws applied with unswerving equality to all, and barking commands to the multitude. Caring for vulnerable children with attention to their individual needs would have led to more flexible rules, and talk based on questions.

These two moral codes, with at least tentative parallels back to the lifestyles of sperm and egg, might also have gained a toehold in the most tenacious of all two-party political systems. In the United States, for most of my memory the hunting party has been the Republicans, symbolically the men, occupying the White House, focused on foreign relations, stalking for GATTs, opening the borders for rambling companies. Domestic affairs have been the domain of the Democrats, in the Congress, symbolically the women. The Republicans tend to encode absolute justice—for example, law and order and hard-line abortion laws. As defenders of equality, the Republican moment of glory came in the racial emancipation of the 1860s. The Democrats excelled in the systems of care developed during the 1930s.

Each party, in fact, chooses to make itself distinct and alluring by constructing parallel poles at both ends of the two-party rung toward those of other binaries, such as international & national. Politics creates an environment rich in binaries. Successful political rhetoric taps into the binaries of the participating psyches and makes parallels with language—"I say yes, my opponent says no"—and a host of other logical fundamentals, such as right & wrong and true & false.

Particularly deep in human evolutionary history is yes & no. By coos of affirmation and snorts of disapproval, elders reinforce the successes and errors of the youths they teach—a pattern that must antedate any accompanying words. Primates condition each other with painful bites and comforting strokes. Reward & punishment imprints yes & no into the neuronal wetware, encouraging certain activities in life's arena, banishing others behind the walls of taboo.

Linguistic and political patterns can spring from those of biology. Look at the term "right-hand man." The metaphor draws a parallel between the right hand's power in action and the increased sphere of personal power possible with the help of loyal and capable followers. Our two hands have been a rich source for parallels. With one hand pointing skyward and the other down to earth, Jesus can map heaven & hell, Buddha can signal the two worlds of his mastery. In Tarot cards justice is right-handed (dexterous) and the devil left-handed (sinister). In their biological division of labor between holding & action, the hands parallel the archetypal binary also displayed as place & motion, matter & energy, form & function, egg & sperm, sphere & tube.

Do parallels, in fact, link many binaries into larger trusses, from gametes to politics, from hands to honorifics? Like poles of a ladder, do binaries trace back and forth across physics and metaphysics, biology and culture? Or is the parallelism all in our heads (to wit, the author's)?

Inside our heads such trusses and ladders are easily hoisted across binaries with alacrity and abandon. When we are standing up, we are generally awake, well, active, and alive; when lying down, we are inactive, sleeping, ill, or even dead. Conceptual patterns being drawn from those biological, we cast the percepts of up & down into a more rarefied realm: by stringing good to up and bad to down. Pleasurable states can then be proclaimed by idioms like "feeling up" and "upbeat," while discomfort is "feeling down" and "low."

As skilled as we humans are at making parallels, we also shake and break them. Though good & bad may usually be borne in parallel with the biological passions toward pleasure & pain, the rungs on the ladder of binaries can be swiveled. Reverse the links and the often avoided pain instead is desired for engendering even greater social gain: Mayan leaders piercing their tongues with stingray spines, Sioux shamans ripping their chests tethered by buffalo bone hooks, Jesus dying upon the bloodied union of horizontal and vertical. Jazz reaches heights when the musicians totally get down.

These swivelings lend a healthy uncertainty and playfulness to the parallels. At not a few points in this chapter a reader may have begun musing, "But on the other hand . . ." That's the essence of critical thinking. How effortless to latch on to an idea, attach a Yes, and then disregard the counterexamples that would happily gobble it up. Banishing potential negations to outside the circle of concern is a seductive and treacherous form of parallelism. Because binding ideas to one side of a figure & ground binary and erecting parallel poles between mental and worldly binaries are such comfortable habits, and because once in place these conceptual ladders develop lives of their own, binaries should be used as spurs to further questioning, not as answers.

Beyond the Binary

It is easy to succumb to and project the simplicity of two. According to evolutionary biologist Lynn Margulis:

> Perhaps because we are divided into two sexes, the human tendency to dichotomize—to divide things into either this or that—is very strong. According to traditional systems of classification, anything alive must be either plant or animal. But taxonomy, or placing organisms into categories, is not just an exercise of science—it promotes a frame of mind that pervades our thinking, colors our values, and affects our actions. Furthermore, that frame of mind may persist even when the classification system becomes obsolete. So it is with the plant/animal legacy.

Margulis sees not two kingdoms of life but five (bacteria, protoctists, animals, fungi, plants). She urges us all to go beyond the binary when thinking about life—from binary to multiplicity. Although a system needs a minimum of two parts, two is clearly no ceiling.

Another way beyond the binary casts aside false dualisms and reveals the original face of unity. In my opinion, the English language leaped ahead when, unlike other European languages, it de-sexed its nouns. Martin Luther King, Jr., effectively smashed much of the black & white projection obscuring the deep genetic and moral unity of all humans. Travel the Möbius strip; the two sides are one.

In an alchemical icon for transcending the binary, a king and queen bathe, animus & anima nude and renewed as one. In another, Maria Prophetissa points to the merging of upward and downward flows from the vessels of earth and sky. Hermes throws his staff of resolution between the fractious snakes. By these means binaries evolve, rather than devolve, into unity.

Interplay between two and one. Artists help us move beyond the binary by challenging our ability to distinguish two from one. Here, alchemical lions and Mexican folk art (made from a coconut shell).

In Hindu scripture Krishna hammered home the transcending of opposites to Arjuna, a leader who stood between facing armies poised for blood. And to give a now-famous sermon, Buddha merely lifted a rose (presumably, words would only have sliced the listener's mind into dualisms). The technique of getting beyond the binary has been highly refined in the Zen tradition; it is sometimes called thinking beyond thinking and nonthinking. Here's how a modern master instructs a novice:

One plus two equals zero.
I don't see how.

Suppose someone gives me an apple. I eat it. Then he gives me two more apples. I eat them. All the apples are gone. So one plus two equals zero.
Hmmmm . . .

In elementary school, they teach that one plus two equals three. In our Zen elementary school, we teach that one plus two equals zero. Which one is correct?
Both.

If you say "both," I say "neither."
Why?

If you say "both," then the spaceship cannot go to the moon. When only one plus two equals three, then it can reach the moon. But if one plus two also equals zero, then on the way the space ship will disappear. So I say, neither is correct.
Then what would be a proper answer?

"Both" is wrong, so I hit you. Also "neither" is wrong, so I hit myself.

These twirls that dance around the lurking grips of binaries go on and on—until enlightenment? Or is that another form of illusion? Regardless, as a space life-scientist in some of my work, I find "one plus two equals three" rather handy; and furthermore, three is a lovely way to go beyond two.

Three is a popular end in the creation of systems. Compared to two, three is more complex—just a bit more complex (one more component) and a lot (at least thrice the internal relations). We process colors in our retinas with a trio of red, blue, and green receptor cones. Kettles and stools must have a minimum of three legs. We have worshipped some heavy trios: look at alchemy's corpus, animus, and spiritus; Christianity's Father, Son, and Holy Spirit; Hinduism's Brahma, Vishnu, and Shiva; rock's Baker, Bruce, and Clapton; comedy's Curly, Larry, and Moe.

Through the galleries of the pantheons most ancient parade trios of goddesses galore: the Erinyes, or Furies; the Gorgons; the Sirens; the Hesperides, guardians of the golden apples in the sunset; the Graces; the Moirae, or goddesses of fate (Clotho who spins the thread of life, Lachesis who measures it out, and Atropos who snips it). This great tradition of triple goddesses has been nearly forgotten by our binary-entranced culture.

To the gridlock of red and blue politics, we can add green. One step out of the prison of feeling and thinking is that of willing. To animals and plants, add at minimum the bacteria. To space and time, add mind. Of course, trinaries preclude some of our most comfortable artifices of

Beyond the binary with threes. The Graces, the Erinyes (Furies), and a three-headed salamander of rejuvenation hint at the plethora of triple goddesses that inhabited the minds of the ancients.

argument: "On the other hand . . ." Fours and mores go beyond the binary, too—but that's for the chapter on layers.

For still much of thinking, we accept our partnership with the metapattern of two. Trying to wrench survival concepts, we draw from the founts of our senses. Through them flow the kisses of bellies male & female, of elements earth & sky, of times day & night. Creation myths postulate pairs, crystallized as two snakes pulled apart by the Cambodian asperas, as waters parted by the Hebrew YHWH. Language is learned as binary minuets of logic.

We may yoke male & female to the sky & earth and sun & moon. But as the Egyptians taught, the sky can be female and the earth male. Nut (pronounced "newt"), who births the sun at dawn and swallows it at dusk, arches above Geb while making love to him. The ancient Japanese also worshipped not a solar male, but Amaterasu, a golden solar goddess.

A rather jolting crossing of parallels comes at the end of the *Mahabharata*. Long after the defeat of his cheating cousins (the Kauravas), Yudhishthira, now old and journey-wearied, climbs the ladder to the sky at Earth's end. Is he greeted by members of his own virtuous deceased family, there in the iridescence? No. Instead, the Kauravas.

We ourselves are poised for reversals in the future, for we have yoked the good to increasing materialism. Stocks must rise, so must the lengths of roads and sales totals, GNP ever upward—the insistent chants of such rallying cries blinds us to Earth's degradation, to our countervailing race into some downhill place.

They writhe across the globe, the binary epidemics: Black & White, Protestant & Catholic, Jew & Arab, Hindu & Muslim, Shiite & Suni, Tutsi & Hutu, and countless more that are capable of rising in an instant from the ashes at the borders of past hates. They take hold of us and command, like viruses do the genetic systems of cells, the futures of good & bad. Gregory Bateson once woke me to the potential stupors of figure & ground projections by warning, "Don't let anyone type you."

Jokes spin the links. Falling down laughing gives a lift. Laugh until you cry. In his theory of creativity, Arthur Koestler called humor an example of "bisociation." We burst into laughter when we are compelled to perceive a situation in "two self-consistent but incompatible frames of reference at the same time."

Where do generals keep their armies?
In their sleevies.

Dropping the initial *h* from this "haha" of humor yields another type of bisociation: scientific discovery. The "aha" of Archimedes's bathtub epiphany, for example, linked the volumes of his convoluted body with the measurable rise in water. In another "aha" Newton linked the fall of an apple with the celestial sweep of the moon, and thus came the theory of gravity. (To whatever degree such events are apocryphal, it is significant that we retell them as mergings of binaries.)

Experiment and control. For flying into the beyond, thinking in binaries is one sure way to travel. The flasks are depictions of those by which Pasteur discovered the role of airborne microbes in food spoilage. The intact neck of the top flask prevented air from contaminating the sterile meat broth; the broken neck of the bottom flask allowed decay to begin. Such realizations born from binary are not just historic curiosities: experiment and control continue to provide the essential "lift" for science to progress, just as the different curvatures of the upper and lower surfaces of an airplane wing provide lift.

The icons of science are often its equations: Newton's for motion, Boyle's for gases, Maxwell's for electromagnetism, Boltzmann's for entropy, Einstein's for relativity. An equation has two sides (two spheres) joined by the tube of equality. Equal, of course—and yet for utility the two sides must also contain a profound dose of difference. In a Sidney Harris cartoon, a beleaguered scientist, after a yeoman's effort in a gargantuan left side of an equation, chalks in the equal sign, then sighs, "This is the part I always hate." Thus $x = y$ symbolizes the supreme accomplishments of science; performance artist Laurie Anderson's song about letting $x = x$ is a hoot. By making a dynamic balance between equality and difference, equations let us relate entropy to probability, energy to frequency, gas pressure to temperature.

We scientists feed our heads with binary graphs. We lecture and illustrate our "papers" with these visual paeans to the powers of vertical & horizontal. We are trained to take binary as the protocol for investigation: an experiment plus a control. We split the realms of sameness & difference with constants & variables. We further split the variables into independent & dependent—the first, consciously altered between experiment and control; the second, assiduously observed to catch a hint into the mysteries of cause & effect.

Scientists are imbued with binaries, from abstract scribblings to laboratory procedures. Our main way beyond is via the "aha," which often reaches a place above the binary of conflicting choice and yields a decision, a synthesis, or a previously unseen possibility. From Zen to comedy to science, skillful use of the binary leads beyond it.

In fact, passing through any of the doors beyond binary—into oneness, pluralism, or swiveling parallels—begins with the binary to go beyond. Binary is both denied and affirmed. We aim not just to go beyond the binary but also to embrace it as the very way to go beyond.

Picture an airplane wing as a metaphor for how our thoughts by binary fly into ethereal discoveries. With forward motion the wing splits the air from a unity into upper & lower. Because the air's total energy is a seesaw between velocity & pressure energies, the extra curve on the wing's upper side is able to boost the air's velocity above the wing with a simultaneous lessening of its pressure. The net pressure on the wing, positive upward, provides lift. Similarly, in cognition, our binaries form systems of primordial differences that lift our ideas. Behind the wing, above & below flow together once again.

Royalty and sun. Royalty have not uncommonly fancied themselves as kin to the sun. One of the oldest of such is a ruling family in ancient Egypt getting solar boons. Nearer today, Louis XIV emblemized his face as a sun. In this print his virtues radiate from the center.

Centers

![decorative molecule graphic] Geometry of Suns and Kings

One sperm merges with one egg, the chromosomes fuse and the fury of growth and form begins. Rivers of proteins channel building blocks and decisions. Far upstream, near the cell's centroid, lies the source of the chemical patterns for the proteins, the holy grail of modern biology, the genetic code.

Some 350 years ago one such fusion of twenty-three chromosome pairs produced the nubbin of the man who would one day call himself Roi Soleil, the Sun King. In lithographs of the solar system, his cherubic face replaced the sun. Bedecked in solar ballet garb he even danced for the royal theater.

Inside the awesome edifice of Versailles, Louis performed his nightly duties in the bedchamber dedicated to Apollo, the Greek sun god. For less officious consulting or consorting, he and company could parade down the broad steps of the palace and retire into the inviting green mazes off the main axis emanating from an immense fountain of glistening bronze, from whose center burst golden Apollo in full-size chariot with horses galloping and snorting.

Like a tornado Louis stripped power from others, swirling the circle of elites tight to Versailles. As the sun beams light, he beamed commands:

I gave orders to the four Secretaries of State no longer to sign anything whatsoever without speaking to me; likewise to the Controller, and that he should authorize nothing as regards finance without its being registered in a book which must remain with me. . . . The Chancellor received a like order, that is to say, to sign nothing with the seal except by my command.

Copernicanism had become widely held, and it offered a new solar identity for Europe's rulers. But the idea of sun as supreme icon had long preceded western science: ancient Egypt had Akhenaten, the Aztecs had Axayacatl. Appropriating the ball of gold as their own metaphor was a skillful move by these rulers. But consider: might politics have developed differently had our sky two suns?

Buckminster Fuller and two of his favorite shapes. Collaged to the right of Fuller, who spoke at the Chautauqua Institute, New York, in 1979, float two of his key polyhedra. The *icosahedron* (top) is stabilized by the triangulated relations among the twelve spheres, which are all on the periphery. In contrast, the *vector equilibrium* (below) has unstable squares as well as stable triangles in the relations among its twelve peripheral spheres. It achieves stability by way of a thirteenth sphere in the center, which is connected to each of the twelve peripheral spheres.

Bucky Fuller's synergetic geometry is an especially rich ground to search for clues to the metapattern I shall call centers. Born from a genome fusion a century ago and living to see a world in which the glories of European kings had been vaporized or reduced to the cultural equivalent of teddy bears, Fuller gave cathartic performances, in his later years, for audiences worldwide. On stage he juggled his discoveries, both physical and metaphysical, as skillfully as Louis his courtiers. Surrounding Fuller like planets were big plastic models. These spherical cages, these polyvertexia of flexible Dacron nodes and stiff wooden sticks, became systems of small spheres and long tubes, the "somethings" and "interrelationships" of his messages.

Two shapes served as a lesson in the geometry of stability. The first, the icosahedron, has exactly twelve nodes dispersed in space as if upon an invisible sphere. Thirty sticks triangulate these nodes into a latticework, a spherical cage not unlike the slightly more complex buckyballs of carbon that would be discovered after his death. Fuller could bounce this omni-triangulated icosahedron across the stage like a basketball.

The other shape—which he called vector equilibrium—can be made from the icosahedron by performing some surgery. Remove precisely six sticks so that all twelve nodes can be symmetrically expanded. If released at this stage, the cage would collapse into a jumbled heap. But now add a new, thirteenth node—smack in the geometric center. To it attach twelve new sticks that radiate to each of the twelve surface nodes, and presto—stability (and bounceability) is restored.

As with the surgically altered icosahedron, so it was in ancient Tenochtitlan, Europe, China, and Egypt. Kings and queens at the centers of civilizations provided stability. Shifts in their rule could throw society topsy-turvy.

> The cesse of majesty . . . is a massy wheel,
> Fix'd on the summit of the highest mount,
> To whose huge spokes ten thousand lesser things
> Are mortis'd and adjoin'd; which when it falls,
> Each small annexment, petty consequence,
> Attends the boist'rous ruin. Never alone
> Did the king sigh, but with a general groan.
>
> *Hamlet 3.3*

Two Great Nucleations

Fuller's demonstration of transforming shapes points to two ways of organizing things into systems. One way is the path of the icosahedron. A rock is a good physical example of same. In both icosahedron and rock, stability is dispersed. Granted, a rock is not hollow like an icosahedron. Yet it is similar because its mineral bits each contribute rather equally to the whole, like the nodes of an icosahedron.

For such a dispersed system in the social realm, consider a rock-and-roll band during a leaderless jam. Some of my peak experiences have been on guitar, with everyone improvising simultaneously. The song's initial rhythm and melody are submerged, and the minute-to-minute sinuous themes vary as leadership is tossed around. One musician and then another picks up on expressive tidbits, which seem to emerge, grow, and then recede, to be replaced, as if by magic—everyone the leader and no one the leader.

Clearly, not all systems are as egalitarian as icosahedra, rocks, and rock-and-roll jams. The solar system and Louis XIV's *ancien régime* express the second model of things in systems, as present in Fuller's vector equilibrium. This center-stabilized pattern is so basic to so many systems from physics to psychology that I regard it as a metapattern. In this chapter the dispersed system will serve primarily as a contrast for thinking, a counterpoint.

Like grape & moon and tree & hoe, the king & sun binary can be framed as a koan—in this case, for contemplating centered systems. The atom & cell as building blocks in the ladder of cosmic complexity are an equally intriguing pair.

Both atoms and cells have dense central regions upon which we have bestowed the Latin term for kernel, inner part, and nut: nucleus. Even before Ernest Rutherford's decisive experiments in 1911, several speculative theories of the atom had posited the existence of a physical and functional center. The hypothesis of molecular vortices, for example, was essentially correct: "Each atom of matter consists of a nucleus or central physical point enveloped by an elastic atmosphere, which is retained all around it by attraction" (W. J. M. Rankine, 1851).

The nucleus of a living cell was first seen not by theorizing minds but by the eyes of the early microscopists, who witnessed the presence of a prominently dark, rounded body within each cell. Experiments proved the nucleus essential to cell functioning. Later, some cells, such as bacteria, were found not to have nuclei. But all cells do have DNA, free-floating in the case of bacteria. And owing to experiments at Rockefeller University in the 1940s, this ubiquitous DNA, whether free or housed (mostly) in a membrane-bound nucleus, became recognized as the cell's genetic code.

The nuclei of both atoms and cells are like the central node of the vector equilibrium because their absence can collapse the whole. Remove the atomic nucleus and the electrons would fly off into space. Remove an amoeba's nucleus and all activity winds down, as protein synthesis grinds to a halt.

How far, however, is it fruitful to carry the parallelism of nucleated atom and cell? Am I perhaps fancying a homology out of a chance homonym? After all, an atom's chemical functioning is also dependent on a full complement of electrons. And a cell can be killed just as easily by dissolving its outer membrane or interfering with its power-producing mitochondria as by plucking out the nucleus. Is it not a

The nucleated atom. Probably written in the winter of 1910–11, this rough note shows Ernest Rutherford's groping toward a "Theory of structure of atom." He further writes, "Suppose atom consists of + charge *ne* at centre + [meaning, and] - charge as electrons distributed throughout sphere of radius *r*."

matter of subjective judgment, then, as to which piece from perhaps many crucial pieces of a system is crowned as the center?

Yes—if all we have to ride on is the model with Dacron nodes and wooden sticks of the vector equilibrium. Then any part that destabilizes the whole could justifiably be called the center. Or, perhaps better, systems with a multitude of essential parts could be called multi-centered systems, implying the possibility of a dispersed system of centers, and thus going beyond the probably too-rigid binary of dispersed and centered.

On the other hand, there are a number of compelling similarities between the solar system's sun, a nation's ruler, an atom's nucleus, and a cell's DNA (or nucleus). Notably, each is physically rather centrally located. (This is least inherently true for a ruler, but even so, Louis XIV was usually inside one of the palaces near Paris, inside France, and could physically reach the rest of the country with ease because of access to horses and couriers.)

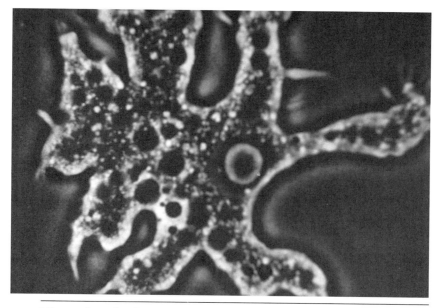

The nucleated cell. Here an amoeba gropes for food, its nucleus a round grape near the center of the photograph. The many dark spheres are vacuoles—sites for digestion, storing water, or accumulating and excreting metabolic wastes.

Second, the center is of unique substance. The atomic nucleus consists of protons and neutrons (the only long-lived members of the type of quark-built particles called hadrons), in contrast to the electrons, which are indivisible particles called leptons. The long, hierarchically folded DNA helix is a chemically distinct molecule in the context of the globular proteins and lipid sheets of the cell. The sun is a glowing, gaseous plasma of fusing hydrogen. And kings and queens—well, royalty are blue blood and at times have claimed divine rights and even an apportionment of holy substance.

Of course the peripheral parts—the electrons, proteins, rocky planets, and peasants—could be said to be of unique substance, too, simply by way of contrast to that of the presumed center. But these peripheral parts are usually numerous, not singular like the center.

Other general properties of centered systems can also be detected. If one describes the chief components of a system—reducing them, say, to a binary or at most several—then the presumed center will surely be on that short list. For example, an atom is a nucleus and an electron cloud. Four primary molecular types in cells are lipids, proteins, carbohydrates, and nucleic acids (DNA, RNA). The solar system is of course the sun and the planets (not, for example, Venus and all else). Societies can be characterized as the rulers and the ruled.

What is more, the center is often particularly resistant to change. Individual electrons are stripped away and replaced in chemical reactions, but the nucleus remains. DNA is assiduously repaired by the cell. Royalty establishes blood-line dynasties and primogeniture for stabilizing wealth and thus power across generations. (Even names were brought into play: the Louis who identified with the sun was, after all, the fourteenth Louis.) The sun burns on despite what the planets do. Clearly the longevity of the conservative center is one key to its providing the stabilizing relations for the whole.

Perhaps the most decisive in this handful of traits that distinguish the center of a system is implied in the line from *Hamlet* quoted earlier. Just replace "king" with "center"—never did the center sigh alone, but with a general groan.

High in the stratosphere a cosmic ray converts a proton in the nucleus of a nitrogen atom into a neutron, thereby transforming the atom into radioactive carbon-14. The atomic nucleus determines the very element. O. T. Avery and colleagues at Rockefeller University raised colonies of pneumococcus bacteria in their search for the agent—the "transforming principle"—that could pass from the dead "smooth," virulent type and enter the living "rough," harmless type, thus transforming the latter into virulence (which depended upon the genes for "smoothness"). They eventually proved the agent was in fact DNA, a molecule that had previously been thought to be perhaps merely a structural filler in the nucleus. Thus "sighs" from the nuclei do indeed echo through the wholes of atoms and cells.

Of all the parts of a system, the center is tied most directly to the nature of the whole. For instance, I can never presume to go to the dean and speak for my department, but the chairperson regularly does exactly that.

Sighs from centers cascade to general groans because they radiate relations to the other parts: an omnidirectional electric field of the atomic nucleus, a network of transfer- and messenger-RNA that distributes the DNA'S patterns, laws and armies from government and the commander-in-chief, light and the grip of gravity from suns. The other parts of a system can be crucial too, but only from the center can we expect to discern direct tubes of relations to all the other parts.

These qualities of centers—spatially centered within the whole, a unique substance, singular, relatively conservative, able to cascade a sigh into a general groan, tangible identity with the whole, and radiating relations to all—might themselves be exigencies or benefits that drove certain systems into centering. Centricity can thus be a design attractor.

Nevertheless, there is a profound difference between the two great nucleations of atoms and cells. As far as we know, atoms did not evolve from a set of extinct predecessors. Atoms simply are. Only in *adaptive* systems—those in which evolutionary processes channel design possibilities toward utility—can we look for wide-ranging patterns (such as spheres and binaries) as design attractors. So it is to life that the search must turn.

 ## Brains, Bees, and Alphas

With nucleated cells the evolution of collections of such cells into bodies could begin. A number of dispersed systems of cells thrive today. The lichen is a symbiotic sheet of fungi and algae, with no single cell or cell type possessing the traits of centerhood. Plants are complex organisms, yet they too seem to lack a center (at least to my eye), with their multiple hormone feedbacks between all organs and their famed totipotentiality—the ability of nonreproductive cells to grow into whole plants. Perhaps the seed is a kind of center in time, a sphere toward whose making all tissues are geared.

The simplest animals—for example, sponges—are also dispersed. But once the path of animal evolution detached its candidates from such substrates as rocks and set them gliding and flying and swimming through fluids, mobility called for a global network of communication and control. A new center was born: the nervous system.

Nerves began simply distributed, as in the nerve net of the freshwater, microscopic (and still attached) hydra. In more complex animals some nerves are aggregated. The balled ganglia of insects and other arthropods punctuate the central nerve channel like beads on a string. The head ganglion is the biggest, usefully positioned near the sensory apparatus of the probing end.

Vertebrates protect this nexus of sensory organs and chief neurons with a nearly solid bulwark of bone. When we reduce a person to a binary, the brain is often one element. We say there is a conflict or convergence between head and heart, between head and gonads, between head and stomach, between brain and brawn, between mind and body.

The brain and its network of nerves, like other centers, is of unique substance. From the viewpoint of cell turnover, nerve cells are the most inert and conservative tissue of all, rarely reproducing. Their longevity provides stability, however, and probably has something to do with the storage of experience and wiring of instincts. On the other hand, nerves are also the most dynamic tissue—in a word, nervous. Neurobiologists call nerve cells irritable and excitable because they can release frisky volleys of ions and drugs that quickly permeate the whole organism. The radiating relations between brain and all else are literally visible in the neuronal networks of fractalized fingers probing to and from muscles, senses, and organs. A bird flickering across my retina triggers me to stop writing, turn, and pick up binoculars—a small sensorial sigh prompts a whole-body groan.

The evolution of the nervous system in some ways parallels that of the genome. In both shifts—from prokaryotic to eukaryotic genomes and from nerve nets to brains—the centers became concentrated and were put inside a formal border. Furthermore, the evolution of bodies

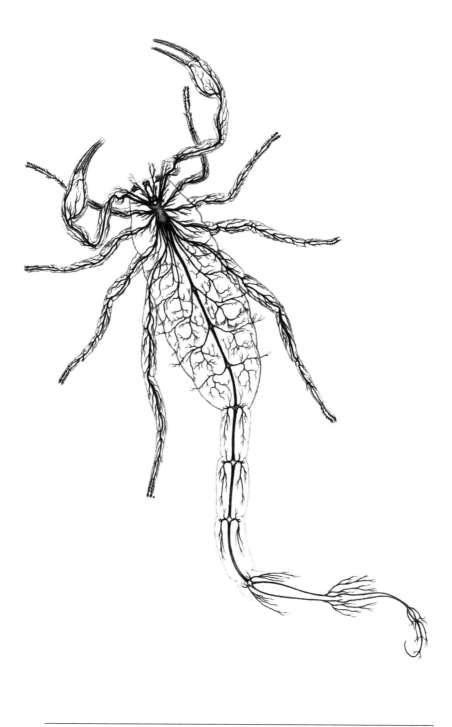

Nerve system of a scorpion.

of cells with identical DNA but differentiated by genes switched on or off apparently became possible only with the enclosing and enlarging of the genome in the eukaryotic cell—resulting in animals, plants, and fungi.

A similar kind of evolutionary step was made possible in what is perhaps a parallel in animal evolution. Nerves allow us to break and re-form behaviors and to accumulate a more-or-less permanent behavioral repertoire. Nerves are to this process—learning—what DNA is to biological evolution. The plasticity of nerves, their ability to change patterns, makes learning possible.

Organisms with nerves as centers made possible another novelty of evolution: social systems. Those organisms with protected, aggregated, and swollen spheres of nerves could begin uniting into larger units, which could move coherently. Flocks of birds, schools of fish, and migrating herds of wildebeest or caribou consist of a large number of nearly identical beings. They are dispersed systems. In a way, they are motile rocks. Within flocks and schools and herds, tubes of behaviors and signals stretch from each individual to a usually close range of neighbors, which stabilizes the whole. The organisms are to their wholes in these social examples as the cells within lichens or sponges are to the whole organism—smaller, centered systems arrayed in larger, dispersed systems.

Social evolution has also yielded examples of the centered way of organizing things into systems. In building bodies into societies, evolution has sometimes been lured by the design attractor of centering. At least two subpaths of social centering can be detected.

First, we see centers in the insect societies of bees, termites, and ants. That centers are key to the patterns of intensely social insects as well as intensely political humans was not lost on early observers: "Though a king in place and power, [the monarch bee] is in sex a female" (Reverend Samuel Purchas in *Theatre of Political Flying Insects*, 1657).

Like other centers, the queen insect is woven from special fabric. In the case of honey bees, she is fed special glandular products to direct her unique development from an egg. The queen is typically huge and can reproduce while the others do not or can not. The fat and fertile queens, bounteously egg-laying, spread their genes into the workers, ranging from hundreds to tens of thousands of daughters. The queen insect is relatively immobile, usually tucked deep into the colony. She is the genetic nucleus around which buzz the electrons of workers. These workers are the architects, harvesters, soldiers, and nursemaids. They arise and die quickly, like proteins of a cell, in contrast to the longevity of the queen, the DNA. The death or removal of Her Majesty (and removal of any incubating eggs that workers could nurture into her replacement), this small sigh in the total biomass, nevertheless would endanger the continuity of the superorganism. The queen embodies the whole like none other in the hive or hill.

My homage to the queen is not meant to slight a dispersed or self-organized component in these intensely social (eusocial) insect societies. Control of the hive's minute-to-minute behavior is dispersed. For example, many shifts in foraging patterns of honey bees come from signals among the workers, whose success or failure at various tasks can raise hive-permeating hormone levels that then bring about changes in the whole. The hive is both centered and dispersed—and

Social centers in ants. Workers of this leafcutter species, *Atta cephalotes*, cluster around their immense mother, the queen ant. The different sizes of workers–all daughters–are various functional classes.
PHOTOGRAPH BY CARL RETTENMEYER, CONNECTICUT STATE MUSEUM OF NATURAL HISTORY.

the binary of centered and dispersed systems is a tool for thinking, not an either-or.

The mammalian branch of life's bush has shown itself also capable of evolving a eusocial centered system—but (known) in only one living twig. African mole-rats, nearly blind and hairless, live underground in large colonies that are uncannily like the colonies of bees and ants. The far more representative path of mammalian social centering produces dominance hierarchies that culminate in what is known as the alpha individual—the leader of the pack, the top of the heap. The alpha individual can be male (as in hamadryas baboons) or female (as in hyenas). Their main mates, depending on the persistence of such arrangements, can also carry the alpha label.

To be sure, the alpha males are often extraordinarily successful at spreading their genes. But even in some species with strong alphas, other males can sneak copulations and testify as competent sperm spreaders too. The alpha female hyena, of course, can only give birth so many times. So dominance is not just a matter of sex and genes. Mammal centers actually have less to do with central gonads and more with central brains. The radiating relations of dominance and leadership from the alphas provide a frame around which the more dispersed relations among others can then in turn coalesce into a stable society. The alpha male chimp decides where to turn and signals the rest to follow. All others can visually center on his actions and know where they are headed.

The general attributes of centricity are reflected in mammalian alphas. As special "substances," alphas are generally big and strong, and they submit to no one. They are singular and (although not the same kind of anchor in physical space as the sun or atomic nucleus) are usually in the thick of things, by roaming the whole and making things happen. A sigh and a glance can send others scurrying with groans. Changes of leadership owing to death and successful challenges create general frenzies as all prepare to adjust.

In some ways the case for the mammalian centers is weaker than for many of the other examples presented thus far. After all, the alpha is not as fundamentally different from the others as is the queen bee from the workers, the sun from the planets, or DNA from proteins. The alpha mammalian center is not a specialized individual, but a role. Perhaps because this role embodies a greater degree of whatever metaphysical substance makes up the network of dominance and cooperation, mammal societies with dominance hierarchies should be considered more dispersed than centered. At the very least, however, they are a step along what is clearly an evolutionary trend toward centering, the prime mammalian manifestation of which is as striking as that of the atom, the cell, and the organism. Out of primal dominance hierarchies have emerged centralized human systems and civilizations.

 ## Government, God, and Gaia

Perhaps Roi Soleil (a.k.a. Louis XIV) would have called himself the Proton King or DNA King had he known about the nuclei of atoms or cells. Sexually he was surely the Alpha Baboon King. Bee King, however, would have been inappropriate, because only a queen could properly place herself with bees. But female centricity of the social insects did not stop the use of bees as models of proper behavior between subject and ruler. In 1722 one observer concluded that the queen bee's power is

> not procured by any Tyranny of cruelty by her exercised over her subjects, but by an innate loyalty natural to these Creatures, not to be diverted by Envy nor Faction, towards this their lawful Sovereign.... Oh, that all the Thousands of this Britannick-Israel were but so loyal to our most gracious King George.... Where Britons, where is your boasted loyalty, that the very insects of our Country shall reprove you?

The icon builders of history may have looked to suns and bee queens to prop the concept of monarchy, but human queens and kings surely have evolutionary roots that color centered societies more with convergence than metaphor. It is easy to imagine a fairly smooth path from alpha baboons and hyenas to tribal chiefs, Nebuchadnezzar, and Cleopatra, from caesars and divine kings and queens to elected presidents and prime ministers. When envoys of King Ferdinand and Queen Isabella of Spain and of other European heads reached the New World, they found alpha individuals topping those state societies, too—Montezuma of the Aztecs, various ahauobs of the Mayan kingdoms, the Inca of the Incans. Remarkably, across theocracies, socialist states, democracies, and single-party corporate states, and despite differences in its extent and means of deployment, power is crystallized around sole bodies. Africa has its dictatorial "doctors," "guides," and

"No. 1 peasants"; Saudi Arabia has its Custodian of the Two Holy Mosques, a.k.a. king; the United States has its Commander in Chief (etymologically related to "chef"), a.k.a. president.

As whole-body movement of physically linked cells drove the development of animal nervous systems, so the need for complex, fast, coordinated behavior channeled groups of human nervous systems into centered social systems. The classic explanations offered by anthropologists for activities trending toward hierarchy are hunting, war (men turning hunting skills on one another), and building great public works. Some of the earliest inscriptions of Mesopotamian kings declare their pride in constructing irrigation projects.

According to Rousseau, through a voluntary yet inevitable "social contract" the general will comes to be concentrated and expressed in a selected few (not necessarily an individual). This unburdens the vast others from global concerns and provides the time they need to concentrate on local tasks of farming, sewing, smelting, teaching. From Rousseau's viewpoint, the center would thus be considered government itself, not the alpha being, but a concentrated group. And what

Government as social center. A village of the Ba Ila people in southeastern Africa, photographed by Mary Light in 1937. Like the nucleus of a cell, the round compound of Chief Mukoblea is distinct and centrally positioned.

an urge toward concentration is manifest! Three hundred million in the United States are rather peaceably governed by a central body (the Congress) of fewer than six hundred. Whether the center be sole or few, Voltaire saw little difference between being eaten by a single tiger or a hundred rats.

Along with individual heads and legislative groups, there is yet another way of viewing the human center. Perhaps those who wear the mantle of government are more like the helper proteins around the DNA. The DNA, the code, is perhaps echoed in human society in the laws of the land. Both DNA and laws, after all, are turned on and off in segments as needs arise. At the heart of societies are the core texts—the Magna Carta, Hammurabi's Code, Ashoka's edicts, the Iroquois Treaty. The framers of the U.S. Constitution, placing more faith in a few tens of kilobytes of weightless code than in kilograms of brains, envisioned the constitution as the organizing source for a social machine that could run by itself, an anchor for the storms that were certain to come from power plays and the passions and fallibilities of mortals.

Which of these three organizing loci is the center of society—the alpha organism, the ruling group, or the source law? Or should all be considered together, as parts of the governing center? These questions may be useful for prodding thought, but they are best left unanswered. Centricity is unquestionably a pattern in the world, but its usefulness as a tool for thinking is easily stanched if it be misused to foster dogmatism.

So too, the center's border becomes fuzzy when scrutinized from disparate standpoints. Arguments can be made that a cell's center is its DNA, or the entire nucleus (when one exists), or the nucleus and all RNA shuttles from nucleus to ribosomes. Similarly, as thinking and acting organisms our centers might well be posited in the brain, but better the entire nervous system. And for the social center, consider the totality of government, with brains in national, regional, and local capitals but also including (and thus equivalent to RNA and neurons) the distantly extending, fractalized fingers of employment: the tax collector, the EPA official, and the backpacking Forest Service agent with rain-mottled map in hand, keeping an eye on the wilderness, days from the nearest road.

Yes, the centers metapattern is a useful tool for thinking about governments, but so too for religion. Paralleling political leaders are spiritual chiefs: big ones, like prophets, popes, and bodhisattvas, and those heading local congregations—the social descendants of tribal medicine chiefs—such as mullahs, rabbis, priests, ministers, gurus, sikhs, roshis, and the Grand Empress of the Chicago Ethiopian Church. Like political congresses, groups can also be centers—from Christian monastic communities in France to their Buddhist counterparts in Tibet—that focus, stabilize, and even create the teachings. And finally, there are key texts—Bible, Koran, Talmud, Sutras, Gita, Dao de Jing—at the heart of the moral, cosmological, and (as in the Koran) economic systems spawned by devotion to a numinous entity or principle.

God itself is somewhat paradoxical. God is often considered the organizer of the universe, sometimes the conductor of events, sometimes the judge of moral existence. Sacred images are centers, too, around which humans congregate—bloody crosses, figures sitting in lotus or floating in the sky, prayer wheels, shrunken heads. When Christians pray "the Lord is my shepherd," they invoke a classic center image. God gets a singular name—Allah, Father, Wakan Tanka, Nothingness, the Void, Abraxas, the Great Goddess, the lapis, that which cannot be named or whose name should not be spoken. God usually has properties in line with many of the generic properties of centers—unique substance, singular, relatively conservative, stable, and long-lasting; God has identity with the whole, it radiates relations to all, and when God sighs . . . Overall, God is a good center candidate.

But does the concept of a god not also offer in some traditions an ultimately dispersed geometry? This is particularly true in the East, in the Dao, in Buddhist and Zen psychology, but also in the more westerly mystical traditions of Sufism, the Kabbalah, and the Christian sacred heart (deemed the dispersed embodiment of Christ). Finally, both ancient and modern pantheisms beautifully fit a dispersed model.

No one thinks (do they?) that their god is centrally located somewhere, such as at the galactic nucleus, in a Deep Space Nine wormhole, or somewhere potentially visible with the Hubble Space Telescope. But even the dispersed concepts have their own level of centricity. When God is pictured as a glowing heart or ball of light in chest or head, the image suggests an organizational center within. Everyone or everything has it. So the total is a dispersed system whose nodes are each small, centered systems. But is God just the sum of the centers, or the entire multicentered dispersed system? Another view takes God exactly as the linked whole, a buildup, which contrasts to a separate soul deep within the body. Instead the soul may be the social

Objects as centers for human contemplation. Humans hungrily attach themselves to the ends of light beams reflected from revered objects. With observers at the circumference, the chosen objects take on the role of center—be they a painting of a dying Jesus, a sculpture of a seated Buddha, a Daoist rock, a cube in Mecca, or something that is revered because it is revered.

PHOTOGRAPH OF MONA LISA BEHIND GLASS, WITH REFLECTED SPECTATORS, BY LYN HUGHES.

body, like mind built from neurons, like an ant superorganism from ants. Thus God is the icosahedron, the system itself. Or if the universe is what we point to, and that includes both centered and dispersed systems, then the universe—the ultimate whole or God—includes both patterns.

When dealing with these paradoxical issues, the established religions offer a variety of answers. A Buddhist holds up a flower. A Christian urges us to set sights outward away from dross matter and by that to get inward to God. A Jew and a Muslim agree, "Of course God cannot be imaged." A Hindu points to Krishna and Radha dancing at the focus of a circle of gorgeous Gopis. A Kabbalist explains, "We place God within the diagram to remind us It is without." An alchemist whispers, "The philosopher's stone, the lapis, is everywhere, even in that turd." A Sioux traces on the sweat lodge floor a set of crossed axes, for the six infinite directions. A Zen roshi exclaims, "You say 'dispersed,' I hit you; you say 'centered,' I hit us both." A Daoist upends and finishes the bottle of wine.

These kaleidoscopic flip-flops may be what excited the Neoplatonists of the Renaissance to make a commonplace of the exclamation that "God is the infinite sphere whose circumference is nowhere and center is everywhere." God is a synergy of dispersed and centered.

Politics is such a synergy, too. For example, early U.S. history is filled with parallel binaries generated by the simultaneous push for centered and dispersed organization. Under the column "centrist" we had Hamilton, Federalists, the core Articles of the Constitution, and the whole nation. And under the column "dispersist" we had Madison, Republicans, the Bill of Rights, and the separate states.

It is therefore of prime importance to see the interplay between the two types of systems as a dynamic binary. For example, in the United States, through the dispersion of power called participatory democracy, voters elect a sole individual.

The interaction of centered and dispersed creates levels. We end up with multigovernments—interacting federal, state, county, and city. We live with multiple masters. We grow up with teachers, playground bullies, parents, coaches, principals, vice-principals, and the neighborhood chess champion; then on to face department heads, division chiefs, shop stewards, supervisors, bosses, bankers, judges, mayors, and the captains of industry. We are in many centers with overlapping ties, large and small. Chuck Berry becomes the "president of rock-and-roll," the *O.E.D.* becomes the "king of dictionaries." Multiple levels of centers are a central theme of society.

Nevertheless, the entire system of centered and dispersed power is in flux. Tension and debates will continue, as one aspect or another demands attention. During the Enlightenment the social abuses inflicted by royalty and an autocratic church led Diderot to proclaim that freedom will come only when the last king is strangled with the entrails of the last priest. In the final paragraph of his final book, Buckminster Fuller urged us to escape from the "misconception that cosmic supremacy is vested in little planet Earth's politicians, priests, generals, and monetary-power wielders."

In the face of a creeping global unification by culture and economics and the first planet-wide manifestations of environmental crises, the pull between centricity and dispersion applies to a greater realm than ever before. Here we might look to biological evolution for prospects. Consider: cells had to encase large amounts of DNA in a bounded nucleus (the evolution of the relatively large eukaryotic cell) before groupings of such cells could differentiate into tissues and start evolving as unities called fungi, plants, and animals. So perhaps the relatively large nation-state, with formalized government and constitution, may be a stage of political evolution necessarily prior to the formation of an as-yet unrealized unity.

Will such earth-scale grouping of human societies follow the animal or the plant pathways shown as possibilities by the evolution of cells into organisms? The animal pathway would imply that one or some of the nations would become a kind of nervous system to the others. We see this today in the Security Council of the United Nations and in the economic pow-wows of a group of seven nations (the G7). This centered theme also, however, is echoed less beneficently in Nazism, in caste structures, in the blood lines of historical royalty, and in the Millennialists of Starhawk's epic about twenty-first-century California.

The plant pathway would indicate a more dispersed role for every nation. Some may be roots, some stems, some leaves, and some even flowers, but none would serve as nervous system, processing information and issuing commands. This is already happening to a degree. Nations seek not rulership of one another but economic exchange. Distinct regions are evolving distinct global functions. Some places specialize in energy production, agriculture, mining, manufacturing, recreation. To counter the inherent pull of education, information, and finance as centralizing agents, advocates of dispersant philosophies should focus on spreading these functions around.

If we tie global politics closely to the climatic and biochemical zones of the planet, the whole of human society can be like Gaia, which is highly dispersed. I take Gaia to be the synergy of biota, atmosphere, marine and fresh waters, soils, and rocks—a whole whose organizational properties we are just beginning to explore. Even if life is visualized as a central sphere that organizes its surrounding spheres of atmo-, hydro-, and pedo-, the web of life itself seems more dispersed than centralized.

The closest ecology gets to a centered system is the concept of keystone species. For example, in the national forest I summer in, cattle are, sadly, the keystone species. In one stretch of the river, with no cattle grazing, I see lush growth of all sorts of trees and shrubs and forbs; in another zone, a wallow of mud, bare earth, struggling and stripped stems, and cow pies. Another viewer might look at both stretches of river and see bacteria at the center of all, as the ecological nucleus ultimately anchoring the electrons of plants, fungi, and animals in stable (or unstable) ecological space.

Nevertheless, much of the aesthetic of nature as a sublime and wondrously tangled web seems to stem from the dramatic contrast it provides to our social world of multiple and often oppressive centers. Nature breathes freedom as an ideal dispersed projection. And it is well that nature is not centered. Imagine what a global empress of mosquitoes could do to us: "Forget the other mammals, get the humans." That's the beauty of nature—no bosses—and it disperses our thoughts on blossoms gliding down a stream.

Some will claim that nature does have a boss, that we, *Homo sapiens*, after nearly four billion years of evolution, are the de facto nervous system of Gaia. We are the planetary brain species.

We certainly do have many generic traits of centers. The use of symbols gives us a special substance. Indeed, we have taken the icosahedron of dispersed nature and molded it into a centered vector equilibrium. Like the models with Dacron nodes and sticks, this act has entailed our removal of a number of original relations, expanding the parts into an unstable, new geometry, then stabilizing them in this position by our own radiating relations to all.

We take the globe as our territory, shifting all to suit our needs. Were we to disappear, virtually every ecosystem would undergo massive resettling. We decide (consciously or not) which species will walk the plank from an ark whose floor space is shrinking, as we reserve more and more space for just a favored few: corn, wheat, rice, barley, pigs, cattle, goats. We irrigate to replace rain, and we are beginning (with controls on ozone-depleting substances) to manage the chemistry of the global atmosphere. Even to leave something on Earth untouched is a decision that must be legislated from the center by edict, law, mandate—and then enforced with gusto.

That is one viewpoint. From another, our very power ensures our doom.

Playing a role in debates about the human role in nature is the binary of centered & dispersed. This binary is deliciously manifest in a dialogue between Dave Foreman, founder of Earth First!, and the epic poet of Mars colonization, Frederick Turner:

> DF: Our environmental problems originate in the hubris of imagining ourselves as the central nervous system or the brain of nature. We're not the brain, we are a cancer on nature.
> FT: If we are a cancer, then I am for it. . . . The nervous system is a glorious cancer that has evolved, and I stand with it. I am that cancer.
> DF: And I am the antibody.

A complaint about centricity that speaks to all times comes from the journal of Mme de Maintenon, the last (and secret) wife of Louis XIV and his greatest love:

> As he is master everywhere, and does exactly what he wishes, he cannot imagine that any one should do otherwise; he believes that if I show no wants, I have none. You know that my rule is to take everything on myself and think for others. Great people, as a rule, are not like that; they never constrain themselves, they never think that others are constrained by them, nor do they feel grateful for it, simply because they are so accustomed to seeing everything done in reference only to themselves that they are no longer struck by it and pay no heed.

Our species has become so accustomed to taking from Earth that we pay it, its resources, and our fellow organisms no heed. Two hundred years ago Diderot—virulently anticentrist toward kings and priests—extolled the centricity that has led to humankind's exploitative habits:

> If one banishes from the face of the earth the thinking and contemplating entity, man, then the sublime and moving spectacle of nature will be but a sad and silent scene. . . . It is only the presence of men that makes the existence of other beings significant. What better plan, then, in writing the history of these beings than to subordinate oneself to this consideration? Why should we not introduce man into our encyclopedia, giving him the same place that he occupies in the universe? Why should we not make him the center of all that is?

Heavy stuff from the chief editor of the core text of the so-called Enlightenment! Similar ideas were expressed by Francis Bacon, who saw the goal of science as taking command over nature. But from the same post-Renaissance era of science and reason that crowned us as nature's ruler (an attitude that can be traced at least as far as the Bible) came yet another aspect of center. What we had to learn to take the reins of nature first required diving deep into its myriad principles, discovering them as centers, as the organizers. Our true friends in high places are the universals within even the infinitesimally humble. An early proponent of these views was Baruch Spinoza, who visualized an immanent God, inseparable from the new discoveries of the natural philosophers.

> The essence [of individual mutable things] is only to be found in fixed and eternal things, and from the laws inscribed in those things as their true codes, according to which all individual things are made and arranged; nay, these individual and mutable things depend so intimately and essentially on these fixed ones that without them they can neither exist nor be conceived.

With such ideas, Europe moved from Machiavelli's *Principe* in 1532 to Newton's *Principia* in 1687.

The principles for which we now search are not always as fixed as the revelations from mathematical physics that inspired Spinoza. The formulations of ecology are fuzzier and inherently tied into messages, communications, evolution, and the interpretations and misinterpretations by organisms. Yet the search for general principles remains at the heart of childhood development, scholarship, science, and wisdom. Out of the age-old vision to make ourselves the center of nature have come the questions (and answers) that now center our attention on nature's infinite details. Not just we as center and nature as periphery, but nature as center and we as periphery.

Our day-to-day practice centers around the known, but our hearts and minds are drawn to the unknown. The unknown empowers our quests. This is the same as God in its mystical manifestation, the same as the best of science.

Now they huddle around a spider, attaching tubes to monitor its carbon dioxide and oxygen exchanges, men and women devoting their lives to the secrets of a spider, harborer of the unknown, of principles that can guide us, both for practical ends and to satisfy the thirst for knowledge. Enough mysteries reside in any one of the millions of species to fill lifetimes of study.

The unknown is a special substance. Long-lasting, influencing all, it is able to swell small sighs into general groans as new facts show us

Nature as center. The spider's web is an evolutionary precursor to the web of science and technology that the human species now weaves.

even deeper levels of new unknowns. We mythologize mystery, internalize it, and hold it dear.

My house is the red earth; it could be the center of the world. I've heard New York, Paris, or Tokyo called the center of the world, but I say it is magnificently humble. You could drive by and miss it. Radio waves can obscure it. Words cannot construct it, for there are some sounds left to sacred wordless form. For instance, that fool crow, picking through trash near the corral, understands the center of the world as greasy scraps of fat. Just ask him. He doesn't have to say that the earth has turned scarlet through fierce belief, after centuries of heartbreak and laughter—he perches on the blue bowl of the sky, and laughs.

Joy Harjo

Dante and Beatrice reach the vision of the rose.

Layers

 ## Vision of the Rose

Following its trip down the fallopian tube, the embryonic sphere of the human nestles into the uterus wall. The mostly hollow sphere flattens, its outer layer eventually becoming the placenta. Inside, cell masses differentiate and jostle in space around several fluid-filled cavities. Beginning at the cavity called the yolk sac (a nomenclature holdover from bird and reptile embryos), and moving inward, one encounters a central disk of cells from which the actual embryo develops. Like a layer cake, this disk contains the three primordial layers, which will become body parts as dissimilar as gut, skull, and skin.

We are born into a world of layers. We dwell within a giant layered sphere. Surrounding the inner core of metals, the viscous mantle slowly heaves, in turn encased by the thin and brittle crust, and then the swirling onionskins of hydrosphere, troposphere, stratosphere, and ionosphere—out of all of which has evolved the biosphere. Within us, too, in the ultrasmall, spheres flaunt layers. Hydrogen's nucleus manages a single shell of electrons; carbon has three and iron eight.

Luminous shells loomed before Dante and Beatrice, in the divine empyrean, the fabled vision of the rose. With their god somewhere in the blinding brilliance of the nucleus, electron shells of angels displayed their glories in concentric zones of a heavenly geometry. This sense of reality as a layered sphere grounded the thinking of medieval and Renaissance Europeans. Giordano Bruno, for example, and other memory wizards often employed levels (on paper: concentric circles) as maps of the physical and conceptual universe. None then knew the layered atom or embryo. But they did know that rock settled below air, and further out were the ever more ethereal and inclusive spheres that held Moon, Mercury, Venus, Sun, Mars, Jupiter, Saturn, and the stars in a grand embrace of Earth.

One day, hiking along the base of towering orange cliffs and poking into the micro-ecosystems of human-scale rocky alcoves, Connie came upon the same picture. I hurried to see: on a shaded wall, faintly visible, was a burnt-red image of concentric circles. Likely painted by the Mogollons between 700 and 1,000 years ago, this symbol—evidently significant to many peoples—has been found throughout the southwestern states. Like all prehistoric pictographs, its meaning can only be guessed. Possibly a sun sign. Possibly what you see with eyes shut at noon in the desert. Possibly coming from a deeper level than

Ancient pictograph in the American Southwest.

the psyche's outer rings of sun and retina—more conceptual, a symbol for a kind of ordering.

Arguably the most important ordering that one will ever contemplate is existence in its totality. From small to large (subatomic particles to the revolving galaxies), or from simple to complex (atoms to the civilizations that brought us Adam & Eve and Atman), the inclusive system of all things has become a modern vision of the rose. One who did contemplate this great flower of empirical diversity was the scientist, humanist, and ascended man, Jacob Bronowski. He wrote:

> Nature works by steps. The atoms form molecules, the molecules form bases, the bases direct the formation of amino acids, the amino acids form proteins, and proteins work in cells. The cells make up first of all the simple animals, and then sophisticated ones. . . . The stable units that compose one level or stratum are the raw material for . . . the climbing of a ladder from simple to complex by steps, each of which is stable in itself.

Bronowski considered this building up of order so fundamental that it deserved a name. With none at hand he proffered his own choice: *stratified stability*. And I have a choice for its icon: concentric circles, as in the vision of the rose and the desert pictograph.

The rise of complexity through layering—through stratified stability—is seen everywhere. A symbiosis of prokaryotes created eukaryotes. Around the gymnosperm's "naked seed" evolved the additional layer of the angiosperm's "covered seed." Solitary insects came before the hive and the hill. Computer programmers ensure that each subroutine checks before assembling the whole. Fax machines incorporated prior stable units: telephone, microchip, printing, paper, and alphabet. Words preceded sentences, which preceded books. All are composites, layer stacked upon stabilizing layer.

How existence is manifest in layers is a significant research front today. Physicists ponder the scale switches from quantum to classical mechanics. Europeans wonder what will happen to their countries with continental unity. Biologists debate the evolutionary roles played by genes, organisms, clades, biomes, and perhaps Gaia. Logicians use levels to decipher the self-referential paradox in "This sentence is false." Like bees to nectar, the economists, linguists, mathematicians, biologists, and computer scientists now pioneering the science of artificial life are drawn to the rose in the fields of nonlinear dynamics, self-organization, and complexity. The emergence of wholes from parts exudes an aroma of mystery, of elusive but profound truths.

 ## Hierarchies and Holarchies

Further upstream in the desert canyon, at a vertical wall beneath an overhang, we found another ancient mural: the symmetrical sides of a red, stepped pyramid. A bird perched upon the narrow top layer. Over all, an arch (perhaps a rainbow? the sky?). Anthropologists believe that throughout the centuries traders from the peoples of the far south—the site of splendors like Tenochtitlan—had journeyed north to this region. Thus it is tempting to read in this mural the influence of these distant peoples' massive stone pyramids, with steps known to represent levels of their cosmos. This vision of the pyramid is a second candidate icon for stratified stability.

Egypt, famed for its own pyramids, also gave us sacred writing, or hieroglyphics. From the same root came *hierarchy*, sacred rule. A pyramid of networks from the pinnacle diffusing downward and spreading horizontally to a broad base is a common image for hierarchy.

Pyramid shapes are typical in diagrams of systems with layers. For example, ecologists distinguish trophic levels in the energy transfers of food webs, from primary producers at the base to herbivores in the middle and carnivores at the summit. Because the mass of each layer in a pyramid, for reasons of structural stability, must decrease with height, so with the biomass of the trophic pyramid. Mice are common, owls rare. The same structure holds for the traditional pyramid of economics, with the large mass of workers grading to the few billionaires at the upper crust.

In these pyramids the entities that occupy a given level neither contain nor are contained by occupants on levels below or above. Owls eat mice, but they do not contain the living mouse populations. Executives instruct their charges, who yet always retain their private, autonomous selves.

Many examples of hierarchies are related to tubes that branch between few and many. Tubes for transport of matter and energy, for example, form hierarchies in river systems, telephone and electrical nets, blood vessels, nerves, and lung passageways. Hierarchical systems of tubes provide structure, as in bridges, the Eiffel Tower, feathers, and arms spreading to fingers. Tubes for information reside in the relations that lead from single DNA to many copies of RNA to even more copies of enzymes to make countless structural proteins, with each step increasing the numbers, as a pyramid widens toward its base.

Arthur Koestler called attention to a general class of hierarchies with an asymmetry in the tubes of relations. In these, information moves up, control down. From electrical control systems in power plants to benchtop science experiments, this flow binary requires two

Hierarchies. The National Science Foundation is a common type. Berries' stems connect with larger stems that lead to even larger stems. In the background is a striking blue Islamic mosaic of the levels of heavens, a hierarchy of consciousness, of steps and points; the higher the level, the fewer who reach it.

distinct types of equipment: input filters for detecting and converging information and output triggers for deciding and dispersing control. These opposing flows of information and control also establish the pyramids of bureaucracies.

Quite different from these pyramidal hierarchies is the pattern of concentric circles, which I will take to mean physical subsumption of the inner layers by the outer. Now I cannot claim that the crisp division I will strike here between the pyramid and the rose, between stacked slabs and concentric spheres, is free of ambiguity. Since both diagrams are maps of systems and not the systems themselves, which map one chooses as metaphor or icon for a particular case ought to hang on the criteria for recasting the system's parts into labels with assigned sites in the icon. Either diagram will sometimes work. For example, the concentric spheres can also be drawn as a pyramid, because the more inclusive levels are inherently fewer in number—the body to its organs numerically fits the pyramid pattern of few-to-many. This one similarity, however, belies a profound difference between the two visions: A body physically includes its organs; the manager of a music shop does not. I will hold the rose and the pyramid as separate symbols—a visual binary evocative of contrasts in containment.

This difference in containment of layered systems is a key point in the systems thinking of ecologist Tim Allen. He has used the evocative and explicit terms *nested hierarchies* and *non-nested hierarchies* to note differences in containment of layered systems. Defining one by negating the other shows that the distinction is fundamental, but this is not the most satisfying method of naming. I therefore propose a distinct name for each.

Let's keep *hierarchy* for the noncontained variety, the command pyramid of generals and their troops. But the contained form of layered system—the very kind that effuses the great mysteries of emergence—deserves its own label. Making it simply another form of *hierarchy* would burden it with the baggage of (usually negative) connotations attached to ecclesiastic, governmental, and corporate bureaucracies. (Don't elevate the hire-archy, the hire-and-fire-archy, the sire-archy, because sire-I'm-tired-archy.) Using a word coined by Koestler, let the nested pattern of parts in wholes that are in turn parts of still greater wholes be called a *holarchy*. Its image is not the pyramid, but the concentric spheres.

Besides their definitional differences in containment, holarchies and hierarchies have other significant contrasts. Notably, holarchies are more obvious, and, paradoxically, more ambiguous, and, ultimately, more mysterious.

Typically the parts of a whole can be discovered quite easily—just look inside and slap on some names. The look may need sophisticated machinery or lab techniques, but the slap is almost instinctive.

Holarchies. Worlds within worlds are here represented by two subsystems within planet Earth: a complex, but delicate, columbine flower and a transparent, single-celled ciliate that is full of organelles. Ancient rock art hinting at the layers metapattern are (*clockwise starting at upper left*) from Panama, Utah, the British Isles, and Maryland.

Holarchies emerge from language processes, our naming gestalts that ring larger entities around those smaller. A hierarchy, on the other hand, gets into the nitty-gritty of what in the world the entities within a level have to do with one another. That effort can demand lots and lots of sweat and may end in frustration. Where in the hell is the Higgs boson—that predicted, still out-of-reach, and so-called god particle of high-energy physics?

The fact that holarchies are so easy to delimit, which leads to their great generality, gives them ultimately a profound ambiguity. Consider: a hierarchy can't really be recognized until its parts and their connections are established. But in a holarchy, the whole is in one's face even if the parts are indisposed to make one's acquaintance. The finding and naming of parts within a holarchy can begin to float into alternating seas of possibilities. A variety of pictures of the rose may all be on the table.

Are we made of organs, or metabolic systems, or cells, or atoms, or memories, or passions, or all the above? During the drafting of the United States Constitution, vociferous delegates debated whether to begin with "We the states ..." or "We the people ..." The people won, yet the country is the U.S. and not the U.P. Gregory Bateson wrote, "The division of the perceived universe into parts and wholes is convenient and may be necessary, but no necessity determines how it shall be done."

When Polonius sought entry into Hamlet's head by asking, "What do you read, my lord?," he was answered with "words, words, words." Unleveled by Hamlet's words, Polonius had to maintain his cool, for although he knew Hamlet was not on the level, he had to try to find out from what level the words came. Hamlet could instead have quipped "morphemes," "verb phrases," "tropes," or "phonemes"—even "letters," for everyone knew he was a man of letters.

Why does this ambiguity of levels within holarchies matter? It is less important with, say, a mechanical system of discrete pieces, such as a bicycle. But with living or quasi-living systems the number of simultaneous levels becomes crucial to what we make of them. For example, ecologists Tim Allen and Tom Hoekstra have shown that organisms are at once contained in a number of equally valid, only partially inclusive, and usually overlapping greater wholes, all of which are crucial to ecologists—populations, ecosystems, landscapes, and biomes. So how does nature work? I am faced with similar ambiguities in deciding how to partition multileveled biological and physical processes in the course of creating a mathematical model, such as a model of wheat growth or a model of the global carbon cycle.

Holarchies are thus deeply and richly mysterious. In contrast, the inhabitants of a hierarchy, a binary, or any sphere-and-tube system share an equality (often of logical type): people in a hire-archy, organisms in a trophic pyramid. We handily envision the passing of electricity, paper, and dollars, or of energy-rich organic matter between relatively equal parts (people) in societies or (organisms) in ecosystems. But what is passed between *levels* in a *holarchy*?

Holarchies evoke those magic words: emergence, holism. They are the links between atoms, galaxies, and minds. How does the whole connect to the parts? What is a cell to the chicken? What is the yard to a chicken? Grain—a component of the chicken yard—excites a chicken's neurons, but does the chicken yard itself do so? The tubes between spheres within a common level in the holarchy drive the transport trio of mass, energy, and messages; cells of chickens relate to other cells. But what type of tube passes between cell and chicken?

Consider a bicycle. Because it is simpler than that black hole for ideas called biology, we might cycle the same line of inquiry. The front wheel is connected to the frame bone connected to the pedal bone connected to the drive chain connected to the rear wheel. But how does the front wheel "connect" with the bicycle as a whole? Through function. The front wheel assists in moving and stabilizing the whole, and it parts company with the rear wheel in its steering function. The connection between a part and whole may thus best be perceived by looking for function (in a broad sense). Whatever one chooses to call it, a tube between part and whole is logically different from a tube between parts on the same level of a holarchy.

Somehow the fingers of fractalized functions interdigitate back and forth across big and small, one and all. Healthy cells build a healthy body; and the body shunts cells around, gnaws on food, and puts itself to sleep. Or is it the heart, the teeth, the brain that make such actions? Complex inward and outward causations loop across layers, relations much more difficult to conceive and describe than the relatively more tractable internal rules, laws, and relations among the parts *within* each level, which range from the elemental forces of physics to the fads of fashion.

Fundamental enigmas notwithstanding, the existence of the holarchy itself leads to a practical pursuit. We poke inside at all levels—

that's basically what science does. A bird may not "know" its cells. But we scientists want to. And we, as the great holarchic level of cumulative humanity, go after the messages. We stick instruments into organs, past cell membranes, into the cell nucleus, where we tease apart DNA, break open molecules, insert atoms.

We go outward, too. And those layers we visualize we also become; each layer exists simultaneously as reality and as conception. We as individuals can nurture relations with the ineffably small, the mind-blowingly big, and the yet-to-be-recognized layers in between. Anthropologist Mary Catherine Bateson raises a call to unleash affections onto a new layer, to embrace the biosphere as our "significant other."

 ## Holons and Clonons

Enough of those heady issues. Let's just plunge into the layers of this universe with our senses. Sit back and I'll give you a running commentary as I replay the video of the classic cinematic voyage through the rose of totality, *The Powers of Ten.* The film starts with a couple picnicking near the shoreline of Chicago. The viewpoint recedes into the sky, paced so that every ten seconds the square of vision enlarges by ten times (its area, a hundred times). Sequentially, Chicago, the weather of the midwestern states, Earth, then Earth and Moon appear at the edges and disappear into the center. Next we see the entire solar system, but soon the sun diminishes until finally the nearest stars are bypassed, blank regions of space linger on, then all stars coalesce when, suddenly, we emerge from our galactic whirlpool of stars, only for this also to fade into a distant campfire. The telescopic ride stops just this side of infinity. What a visceral rush and kick to one's philosophical base!

Yet compared to what comes next in the film, this outward journey through the rose will come to look blandly repetitive: stars within clusters within galaxies within clusters and superclusters and from far enough away they all look just the same, bright bulbs in the heavens. These various scales of fractalized sameness, ruled by gravity's grip, do contain some grand enigmas in local features, like neutron stars and brown dwarfs and cosmic bubbles and superstrings (spheres and tubes?). However, more mystery lies within a mosquito than all told in the astrophysical universe.

Like a roller coaster over the hump, *Powers of Ten* next falls back to Earth at five times the speed of the outward voyage. Stopping briefly above the couple, it then begins to descend at the original pace, inward to a hand, through the skin cells, into a capillary, and then entering a lymphocyte of the blood, toward the DNA. Biology has distinct layers and its wholes have parts related by functions, bound by messages. That is its fascination.

Now astronomy holds coordinated systems of unique forms too, if split finely enough. But for the most part, stars are switchable; galaxies are too, so are cosmic bubbles. Not so, switching thumb and little finger or bases of DNA. (Even switching the role of those picnickers of the 1960s, her into providing family income and him into child care, would probably have been rough.)

I would like to take this test of switchability between parts within a whole—their interchangeability, exchangeability, and replaceability with often not just a single other, but any of many available—as a primordial distinction. But before seeing where it leads, the distinction could use a moniker. The repetitive things have been called parts of homogeneous systems, aggregates, and clusters, as contrasted to members of heterogeneous systems. I would rather focus on the parts themselves and find terms for those. I want names that call out the sameness between two stars and the difference between wing and leg. I suggest the terms *clonons* and *holons*.

Clonons, at the far end of the spectrum progressing from different to same, are literal in the clones of genetics—in aphid clusters, for example, with identical, asexually reproduced DNA. Clonons—bricks in a wall, red blood cells in a capillary—perhaps epitomize parts that sum to their wholes. Switch numbers in a sum and the sum remains the same. This is because each number is an aggregation of ones (or parts of ones), those conceptually ultimate clonons.

Clonons form the basis for much of the current and exciting work in complexity theory. Identical units interacting in accordance with simple sets of behavioral rules fascinate because, in these clonon systems, wholes emerge as more than the sum of parts. With computers as the tool of choice, researchers have composed simple algorithmic birds, ants, and wasps that nevertheless aggregate into flocks and colonies; these marvelously negotiate obstacles and excite our admiration in their economies of food-finding. Grains of sand falling and

Clonons. The biological urge to reproduce masses of others is here represented by a sinuous strand of bullfrog eggs. We nurture that urge in vast agricultural fields, in order to feed the masses that form movements—of consumers, of voters, of protesters (here in Russia). On the placard are collaged the atoms (from an earlier figure), signaling clonons at all levels.

triggering avalanches upon sand hills, fires igniting and consuming squares of trinitron forest that afterward regrow have led to such arcane-sounding concepts as "self-organized criticality," which has now entered the lexicon of complexity theorists. The mother model of all these, and still a potent research tool, is the grid of cellular automata on a computer screen, which flickers with spots of on and off, clonons all.

Complexity investigators have used systems more complex than identical cells with identical rules of relations. Nature has, too. One could have spheres and cubes, for instance, with separate sets of rules. In nature, this pattern is found in the organisms with a countable and small number of cell types.

The just-visible, spherical green colonies of *Volvox* and another, much larger species of algae called sea lettuce each have precisely four cell types. Each produces male and female gametes. In *Volvox* the other two types are the numerous small cells for vegetative growth and a fixed number of larger cells that function as "seeds" that can float away and asexually start new colonies. The two nonsexual cell types of sea lettuce form a more spatially functional synergy, one type for the "leaf" and the other for the flared stem holding fast to a rock.

Volvox and sea lettuce therefore contain four holons, the cell types. Here the types themselves are holons; the individual, numerous cells of a certain type constitute clonons. From small to large, a holarchy is built: clonons within holons, these holons within the whole. But what happens next, what is that whole in turn a part (a holon or clonon) of? Where fits that algal ball, that green sheet with tube? What system embraces the sponge (with about ten cell types), the sea anemone (with about twenty cell types), the human (with more than a hundred cell types)?

Before answering, it might be worth stepping out into some theory for a moment. In the night sky—in its purest, fractalized, abstract reduction as gravity-bound units forming larger gravity-bound units, and so forth—clonons build wholes, which in turn are clonons that build larger wholes. Clonons of clonons of clonons are useful in biology, too, as shown by the holarchy of a half dozen layers of clonons physiologists have discovered in our tendons.

In the stars, then, clonons form further clonons. In the *Volvox*, clonons form holons. Now clonons and holons are a binary, conceptually at least. A binary of things hooking up can make four distinct binary combinations. We have two of the permutations so far. What about the other two?

Holons can form further holons. For example, the major functional regions of our brain—cerebellum, cerebral cortex, thalamus, pituitary, pons, and medulla (add or subtract a few, as you wish)—are holons within a whole. The brain, in turn, is a holon with sibling holons such as heart, lungs, liver, and gonads. My PowerBook, with its holons of screen, keyboard, disk drive, hard drive, and microprocessor, is itself a holon. It's the only one I have (and if it breaks down, so does this book).

But I could purchase a replacement computer, so in that sense it is a clonon. It's a clonon in the manufacturer's warehouse and as a unit coming off the assembly line, but a holon with me at home. Can a thing be both clonon and holon?

The psychedelic rock group Pink Floyd insinuates, "All in all, you're just another brick in the wall." Bob Seeger insists in that singing wail of his, "I'm not a number." To a sponge diver with bag to fill and a marine biologist studying reef population changes, a particular living sponge is just a number, its differences with other sponges of the same

Clonons at work, holons at home. Author's grandmother (second in foreground) on line at the Buffalo Bolt Company, 1919, then with daughter, son-in-law, and grandson—here gaping in open-mouthed wonder at his first inkling of metapatterns.

species or at even higher taxa are unimportant. But to the yellow and white striped reef fish, at the moment when it rests in the sheltering nook of that sponge's folds, the sponge is a holon. For many people the same pattern applies: we are unquestionably holons at home, but probably clonons on the job.

The distinction depends on how one navigates the levels of the system, the holarchy. The somewhat arbitrary width of the observer's window is hugely important. I may see a holon where you see a clonon. Basically, the wider the window, the more likely that a thing will shed its holon skin worn at a local level for a clonon cloak at the global. When I justify killing a fly, I invoke its clonon nature, possible so long as I do not observe it so closely that, to me, it becomes a holon.

The assumption of the crisp presence of a holarchy collides with its inherent ambiguities, leading to some fascinating results. Mixing clonon and holon properties at various layers occurs all the time in the model-building of scientists. A leaf is a holon to its supporting twiglet, yet a clonon at the level of all leaves. Leaves as a population form a holon joined with a few other main holons of stems, roots, and flowers—if, that is, one builds toward the tree as unit. But by sticking directly to the criterion of primary photosynthesis, one can form an alternate series of layers of clonons into clonons: chloroplasts in cells in leaves in canopies in forests. Such clonon schemes that bypass the tree's holons (and the tree itself) are used in many biosphere models. Skilled science parses the universe into levels with skips like Hamlet's jump in language.

Clonons and holons are abstract design attractors, often approached by live and not so live things. Yet, the distinction in practice fades in and out, as complex and multilayered effects and causes, like insights glimpsed and then lost, like mirages clear from a distance but never reachable. Despite this elusive fuzziness, considering holons and clonons as idealizations may reveal even more than what follows directly from their definitions. They may bear unique properties that have tangible and far-from-fuzzy effects on the very numbers of each that can be found in systems.

Handfuls of Holons

What's a clonon alone? By definition, it takes at least two to call either a clonon. Two clonons of oxygen make a molecule; two eyes and two gonads follow the parsimony of binary. At its simplest a binary of clonons offers redundancy, the practical backups of engineering. For us two eyes provide depth vision, their separation astride the nose adds some holon spice to their skeletal clonon design. And the two identical atoms in a diatomic molecule must both be present for the molecule to exist, more holon spice.

Engineers at NASA often specify triple backup, lowering the overall failure odds by raising the exponent that acts upon a single mishap's fractional probability. The proverbial three's a crowd ignores the holon portions of people. Many flowers exhibit symmetries of three clonons in circles (especially three's multiple, six). Four and five clonons occur, too. But truly, these low-number systems only put the power of clonons into low gear. Groups of hundreds or even thousands of clonons still only begin to tap the possibilities. Even a million clonons together can be merely a mote in fulfilling their potential for massing.

Grains of sand, the human population bomb, oxygen molecules, leaves—in diverse instances more clonons quite simply do more of what fewer do. Redundancy and the sheer building of mass are key properties of clonons. When the nectar is flowing, lay more eggs. A growing cell slots more lipids into its membranes. (Borders inevitably use clonons because sloughing & resupply is an easy way to cope with entropy.) Markets, religions, and cultural movements thrive on increasing numbers of clonons. Once the universe has discovered the benefits of a particular type and is able to perform the initial replication, the hard part is over. A thousand, a million, and more might very well be even better at doing whatever it was that two were better at doing than one.

In contrast to this tendency of clonon numbers to rocket into the astronomical, for good reasons the number of holons in a system is often held down. How far? Well, for example, here are some systems of holons—how many in each?

U.S. Cabinet departments, Jesus' apostles, types of rooms at home, rows or columns of Periodic Table, cell organelle types, major body systems, bodily senses, teams in sports leagues (clonons to some), parts of brain, systems in a NASA diagram, main characters in *Roseanne*, in *Romeo and Juliet*, Earth's climate zones, Folds in Buddha's Path, Graces of Mary, E.C. members, simultaneous courses taken by a student, Sisters of Oil, chapters in books, articles in *Scientific American*, cake layers, nations in the

G7, chambers of a heart, tribes in the Iroquois nation, terms in the Navier-Stokes equation, directors on corporate boards, nodal stars in a constellation (but *not* planets in this solar system), geophysiological organs of Gaia, big gods in Olympus, ant society castes, top ten music hits, schools in a university, branchings between tree trunk and leaf, chess pieces, insect body parts, Greek Elements, Chinese Elements, families of fundamental particles, orchestral sections, fingers, habits of highly effective people, tribes of Israel, Mars Viking Lander experiments, Moses' Commandments, Provinces in Hin Su's China, biological Kingdoms, spheres in the Kabbalah, fundamental forces of physics, cell types in an organless sponge, boxes in a crude carbon cycle model, atoms in carbon dioxide or water, members of Monty Python's Flying Circus, Sages of the Bamboo Grove.

The answers should range from three to not much more than twenty. After number twelve English switches to the cloning aspect of the teens and beyond. So twelve might be a convenient number to remember for holons. Since holon numbers fall in a range, I like to think of holons in handfuls. Rather than sand on the beach, holons tend to become pebbles in the palm.

What factors might drive (or at least nudge into existence) this pattern of handfuls? Or is this just the author's head in the clouds, a landscape of murky shadows becoming a mind-projection screen? If you are inclined to think the latter, take another look at the list. Exclude, for the moment, the few examples from physics and chemistry. Everything else on it is either a living system, a product of a living system, or the subjective interpretation of a physical system given by a living system (us). The notion of function (either intrinsic to the system or to the system observing the system) is key. Holons may come in handfuls for reasons.

To the extent one buys into all the metapatterns thus far presented, a handful of holons comes as a bonus: assume a border and interior, add a center, and separate the peripheral parts of spheres and tubes into a binary of types, stir in a sheet and a small hierarchy, maybe add one more for the environment around the sphere, and the result—a handful of holons.

Circular reasoning? Perhaps. And was the previous list biased to low-number examples? Of course. Why would I want to mention

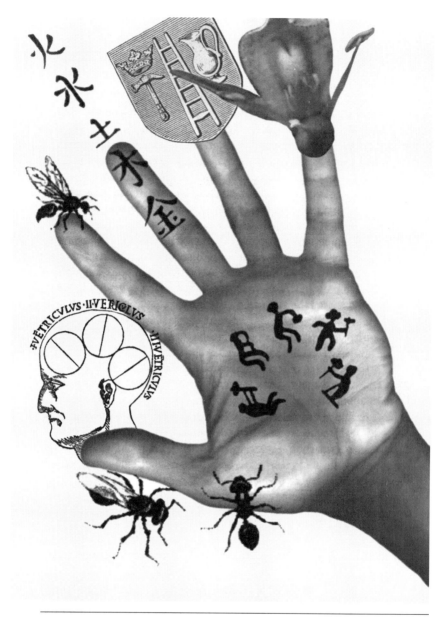

Handfuls of holons. *Counterclockwise from palm:* functionaries in an ancient sun ceremony of Moki Pueblo; flower parts—stamen, petal, sepal, and pistil (perpendicular, with white tip); goldsmiths' banner in medieval France, showing alliances to slaters (ladder) and pewterers; ancient Chinese elements—metal, wood, earth, water, fire (going outward); head with faculties—imagination, thinking, memory (front to back); ant castes—winged female and worker (on thumb) and smaller winged male (on index finger) ready for mating flight.

such theory-destroying examples as the forty functionally distinct areas of the mammalian neocortex, the fifty-five members of the U.N.'s Sustainable Development Commission, the hundred variables of turbine design, the 613 commandments of faith for pious Jews, the thousands of protein types in cells, the hundred thousand genes in the human genome and objects in Internet, Earth's millions of species?

Let me however defer discussing at least some of these counters until the next section on alphabets. Anyway, no claim prevents holons from attaining large numbers, only that no design (read: functional) pressure shoots them into that realm, as does the advantage of repetition to clonons. So perhaps the difference is only relative, perhaps the numbers of holons only comparatively get left in the dust by the clonons rocketing away. Is this all? Or might diverse drivers actively brake increases in the number of holons or even steer them toward clustering in handfuls?

One possibility stems from Bronowski's idea of stratified stability. The parsimony of binary extends to other small numbers. It only takes a few to create new. In the building of stratified stability, new levels of systems require only a few elements as long as these interact synergistically. Simplicity is easily and thus frequently accessed in the trials and errors of physical, biological, and cultural formation. A scissors is two knives with holes and a hinge. Exceptions to two sexes are other low numbers. The European Community starts with agreements among a few nations. A few types of atoms make most molecules.

Furthermore, it may be difficult to wedge newly invented holons into an already functioning sphere of holons born from the parsimony of few. Once created, the new level (a table, say) relates as a unity with others at its level (chairs, people, dishes) and gets stabilized. A table has two holons: flat surface and legs—what more could be added? Expansion leaves for Thanksgiving dinners? Or would leaves be a sub-holon of the flatness holon?

Function-space could hold some universal types; for example, the border holon. One set of elemental functions has been proposed by systems theorist James Miller, in his monumental *Living Systems*. Seven levels of living systems (cells, organs, organisms, groups, organizations, societies, and the emerging supra-national system) each exhibit nineteen functional holons (boundary, ingestor, input transducer, extruder, output transducer, supporter, matter-energy storage, memory, distributor, converter, producer, motor, internal transducer, channel and net, decoder, associator, decider, encoder, and reproducer). All living systems might require metabolic subsystems for border-crossing, storage, transformation, and use of matter, energy, and information—in other words, a handful of holons.

A similar, metabolically driven handful seems present in executive cabinets of governments. Although details differ, they usually reflect rudimentary "living" functions, such as transportation, commerce, defense, energy, interior, agriculture, and foreign affairs. Parts of a green plant make a minimal system: roots uptake water and minerals, leaves absorb solar energy, stems support, and flowers reproduce. Robert Pirsig described the components of his motorcycle as a functional binary: the power & running assemblies.

There is another reason that holons seem to occur in handfuls. Too many cooks spoil the soup. This aphorism captures the possible confusion, the excluding competition, the overlapping jostling for functions if holons were numerous.

Modeling experiments have shown that mathematical food webs usually decline in stability as they get more complex in species. One exception even further proves the rule. When species share strong links in loops smaller than the total system and have only weak links to others in somewhat separate loops, the mathematical holarchy is stabilized. This pattern of middle modules has indeed been found in nature, in the three separate and concurrent detrital nitrogen webs in soil. Stability can come from reducing the number of holons in the level nearest to that of the whole by grouping them in an intermediate layer.

If holons do worm their way into large numbers, to remain so they must hold out against a trend that finds efficiency in recombining elements into an intervening layer of a smaller number of subsuming holons. If too many cooks spoil the soup, some should start the salad.

Recently the manufacturers of White Cloud bathroom tissue decided to fold it into the Charmin line, as Charmin Ultra, because too many names confuse the consumer. Our bodies have hundreds of cells types, but grouped, for example, as a dozen morphologically distinct types of neurons. The patent application for a pistachio huller enumerates hundreds of parts, yet manufacturers would group these into a smaller number of named subassemblies. NASA designs an advanced plant crop habitat with tens of thousands of parts by conceiving a handful of functional modules, each with a handful of submodules. The fifty states are often grouped into the northeast, south, midwest, southwest, northwest, and California. The United Nations configures a layer of voting blocks from the nearly two hundred nations. And, after billions of years of evolution and millions of concurrent species, we come along and simplify the planet into the Cold War's East and West, or now the Environmental Crisis' North and South. These political simplifications and even the exigencies of engineering may be casts from a mold in our brains for low numbers of holons.

What if I had indicated, prior to the details, an intent to elaborate (or even name) thousands of reasons for spheres and more thousands for tubes? You might have closed the book. Or how about a spheres-and-tubes model of the carbon cycle with two hundred unique boxes? Good for a government grant, perhaps, but not for explanation.

Most people can recognize three sides when a triangle is flashed too briefly to allow them to actually count. Flash a quadrilateral? Still no problem to discern four in a single gestalt. A pentagon? Still easy. Then figures with six, seven, eight, nine, ten, eleven sides? Starting to lose it? Do such numbers relate to limits of simultaneous memory?

What if there were a hundred sports teams in a league? Could they remain individuals in the mythic struggles lived by the spectators? Is this why the number of major characters in *Hamlet* is about the same as teams in a league? If a jazz or rock ensemble has more than six or so musicians, groupings are scripted, with similar instruments playing similar parts. Orchestras are in sections. Psychologically, holons are things we distinguish. Our ability to hold holons in mind may derive from a mental parsimony.

Are the number of Secretaries in the President's Cabinet and the number of Apostles in Jesus' inner circle related to innate limits of human information processing? Management theory recommends that one person supervise only about eight others. A landmark paper in applied human psychology is "The magical number seven, plus or minus two." An application is phone numbers, since we can remember a series of numbers of up to seven digits, but memory breaks down rapidly with more. (Area code must be a separate layer, and Connie remembers her fourteen-digit phone card only with the crutch of spatial layout on the touchtone grid). Neurobiologist William Calvin predicts a similarly small handful of storages in the rapid-fire sequences of neurons, which evolved as we coordinated hunting throws. He says this low-number sequencing carries over into why we parse poetry and emit words in small bundles. (Averages in this chapter: 6 letters per word, 11 words per sentence, 7 sentences per paragraph, 20 paragraphs per section, 5 sections in the chapter.)

Low-order counts may be projections of our psychological limitations. But again, our mental numbers may have come from the same processes that built holons into holarchies everywhere. Grouping a few storages would have evolved first as trials, and perhaps such low-number synergy would have taken our brains and therefore minds to a new level. Still, some holons do seem to come in large numbers.

Alphabetic Holarchies

You know like how DNA has these things called repeating units, introns, or junk or something? They're uh not yet well understood at all even. Like they could be harmless selfish chinks of code, sort of like some hidden computer viruses that you don't even like know are there. Or um these enigmatic clonons could be something really important of like you know the whole thing or something.

Whatever the introns, they get excised after the transfer of pattern from DNA to RNA. The edited remains, the exons, large pieces containing many scores of triplets, are then stitched together to form genes, which some biologists claim serve mostly as heuristics in these molecular mysteries. Other biologists recite the rule of one gene, one protein. So a number of strata appear in genetics: four bases build sixty-four possible triplets (for unknown reasons only forty-eight occur); these triplets build introns and exons; these exons (possibly as few as several thousand) comprise gene subunits to construct a hundred thousand human genes in a world with enough unique genomes to reproductively isolate and therefore define perhaps ten million living species.

Since single bases and single genes can affect bodies, and single species the world, the exons and genes and genomes are all holons in huge numbers, based on an ultimate handful in a combinatorial buildup of stratified stability. This holarchy of expanding holons through creative recombinations gives DNA a close partnership with language. Its twenty-something letters build a similar holarchy, with thousands of words in daily conversation (Shakespeare used about thirty thousand), to an infinitude of sentences, and then stories galore.

The DNA holarchy from bases to genomes seems to fit the group of properties laid out by Canadian physicist and linguist Robert Logan for generic alphabets. They are: (1) elements are relatively few; (2) each element is atomic, that is, repeatable; (3) a huge number of seemingly complete sets (words) form from these; and (4) variable ordering is key.

In the language of this chapter, these properties can be recast: generic alphabets consist of a handful of holons; these holons are types of clonons; the clonons are not merely massed, but—and upon which the properties of whole depend—are ordered in particular mixed groups; these groups can themselves be combined by permutations, creating ever-enlarging possibilities for ever enlarging.

One alphabetic holarchy we use every day yet hardly think of as such. With a handful of digits (ten: 0–9, plus a dot) and their relations in the positions of tens, numbers huge and infinitesimal are made. These numerical words in turn are strung into sentences by addition, subtraction, integration, exponentiation, and other operators. In my technical papers I usually keep equations to a minimum, about a paragraph or so. Full-time mathematicians, however, compose stories and epic poems out of these numerical units.

The T, A, G, C bases of DNA are clearly at the base of an alphabetic holarchy. There may be intermediate layers, too, in the sequence of DNA to RNA to amino acids, which makes the genome system more of a hierarchy—non-nested, noncontaining—and in which increasing numbers of clonons are evident at each step. Nonetheless, once the pattern is transferred by hierarchy between DNA triplets to amino acids, another alphabetic holarchy is entered. The twenty-three amino acids (like letters) in groups of hundreds (and more) form proteins. These proteins, countless like words, in actuality occur as a few thousand types in medium-complex cells, approximately the number of words in educated human speech.

Physics, too, has been heralded as having more than a soupçon of similitude to language. Its ordering of parts is not linear, like the scripts of genes and memes. But one would not expect the rules of functional language to be precisely like those of physics and chemistry—which just are what they are. Instead, self-patterning of electrons makes the whole of their numbers more than their sum, as en masse they push any additions into new quantum harmonic shells. The atomic alphabetic holarchy builds from few to many types. Two types of quarks make protons and neutrons, which, as nuclei with electrons, make about a hundred elements. These hundred elements are, in turn, the "letters" that throw the possibilities in the next layer, the "words," or molecules, toward infinity.

French physicist and philosopher Hubert Reeves believes this alphabet-like way of building complexity is so fundamental that,

Alphabetic holarchies. Spiraling outward from the center of the shell of a desert land snail are six quarks; two triplets of "up" and "down" quarks; the three constituents of atoms; four atoms; three simple molecules; one geochemical equation (the weathering reaction of silicate rocks); a string of DNA base pairs (A, T, C, G); three-letter abbreviations of amino acids; two enzymes (proteins); the virus whose DNA was the first to be completely sequenced; the scientific name for wheat; the first six letters of the Russian alphabet; the founding motto of the Royal Society.

rather than saying "nature is like language," we should spin the binary and say "language is like nature." The increasing possibilities from "words spelled out by the component letters in nature's organized systems" drives a "scale of complexity." Bronowski would concur.

Many other holarchies exhibit some of the same complexity-generating layers as do the alphabetic holarchies. With three pairs of shorts and ten T-shirts, I get thirty different summer outfits. But these are weak holons compared to the essential roles of each of a hundred thousand genes.

Another kind of alphabet-resembling holarchy occurs when several holons can be mixed nonatomically along a continuum of amounts. What about the way we combine foods and spices in recipes? mix three primary colors into all colors? watch plays with skillfully mixed sequences of a handful of characters? engineer water, electrical, and control systems with a modeler's tool kit of mathematical capacitors, resistors, and so on? Religions stir a few memorable moments and archetypes into a heady stew of beliefs.

A further wrinkle in generalizations about alphabetic holarchies appears, as combinations on new levels exhibit not just increasing possibilities but new levels of simplifications—types within types. Each of us has a hundred thousand genes, yet they are turned on and off in patterns that make only several hundred types of cells. Proteins are not just amino acids; they contain intermediate modules, such as alpha-helixes and beta-sheets. Sentences contain patterns of logic that constrain and direct the placement of words. The infinity of number groupings is actually simplified by algebra (literally, the science of reuniting, the reunification); numbers are generalized by single letters: $a(bc) = (ab)c$.

All in all, alphabetic holarchies are central, conceptual suns around which spin many related and partially overlapping issues, such as massing versus mixing, unique ordering versus higgledy-piggledy fusion, and alphabetic holons whose examples are strict clonons (all letters A) versus those whose examples share only certain properties (all nouns). How many alphabetic holarchies are there? Three big ones—physico-chemistry, biology, and culture? More? Less? Only one, since genes and ink are ultimately made of atoms? Or just the one in our heads? Robert Logan argues that the presence of an alphabet in our heads has had profound import in our portrayal of the world: "As scientists pursued their ideas in their particular fields, the alphabet served as a model for organization. The fact that they consistently use the letters of the alphabet to designate their categories is one hint of this influence."

Chemists use letters singly or in pairs for the elements, and thus write molecules as words. Physicists prefer the letters of ancient Greece for the types of radiation. Mathematicians employ letters both ancient and modern (they need all they can get) for their derivations. Biologists nickname genes, such a *mec-4*, *deg-1*, and *wg* (for *wingless*). Geologists investigate the impact crater at the extinction boundary called the K–T. Teachers give letter grades to students from the elementary to the graduate levels. But beyond names, the search for causes and effects (for example, forces and accelerations) might be traceable to the alphabet. So is the industrial revolution, with "repeatable uniform elements of gears, screws, and levers in linear configurations to achieve specific mechanical tasks."

Fluid engineers see flows as infinite admixtures of friction, viscosity, drag, velocity, and degrees of turbulence. Paleoclimatologists place their imaginations into Earth's past climates by envisioning different carbon dioxide levels, configurations of continents, orbital patterns of Earth, and solar output. Physics has its four forces; the ancient Greeks had four elements, and the ancient Chinese five. Carl Jung saw people as words from an alphabet of four personality types; psychotherapist Karen Horney saw personal problems as words from an alphabet of three root conflicts.

My colleagues and I often partition wheat growth into an energy cascade of three or four processes, knowing perfectly well that we are ignoring the complex dynamics of thousands of proteins. Yet we are confident that some aspect of truth will be revealed to us in this simpler division. Pondering graphs of the energy cascade, we try to interpret the state of the whole, and what part each process plays in relation to the others as growth ensues.

Humans invent language and then use language to reinvent the world. We project our alphabetic minds through language as a prototype by which we reconnoiter, test, and discover. A prime subpattern of the metapattern of layers is thus the alphabetic holarchy, with holons of clonons melded in local geometries, layer upon layer—small numbers at the core and then ever increasing with the layers. This metaphor has been fruitful. Language is indeed like nature; nature is indeed like language.

The letters x, y, and z, in addition to their mapping the three dimensions in space, are also the traditional designates for the unknowns during the discovery phases of science. Such notation more than hints at the possibility of deciphering the words that hold these x's, y's, and z's. The unknown is veiled in a code—a mystery with tantalizing clues as to how we may focus frames for larger pictures, decipher the elemental families of spheres and tubes in holarchies, hear tongues in trees, read books in running brooks.

Galileo, speaking as the sagelike Saggiatore, mused:

> Philosophy is written in that vast book which stands forever open before our eyes, I mean the universe; but it cannot be read until we have learned the language and become familiar with the characters in which it is written.

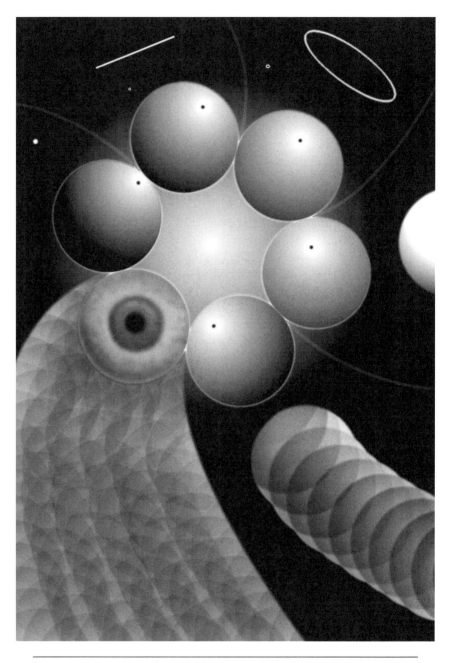

Dance of the dimensions. The spot on the spinning sphere is seen and tracked by the mind's eye, thus giving birth to subjective time. Time is envisioned as the fourth dimension in a holarchy of dimensions, which begins with a point (zero-D) and ends with mind (five-D).

Calendars

Space, Time, Movement,
Memory, and Mind

Imagine you are floating in a space without end. A dim, translucent light fills the void. You watch, but no movement breaks the stillness. Nothing happens, nothing changes.

Now, some distance before you, a disk appears, a solitary roundness in the void. Your attention rises, then wanes. For again, nothing happens; nothing changes. The disk is there, that is all. But wait. Something is happening. A spot appears along one edge and glides across the face, disappears, and then returns. The disk is, rather, a sphere. And it is the sphere that is moving, not the spot. The sphere is rotating in space.

That spot becomes the most precious thing to you in the universe. It gives you something to watch, something to do. You can count the rotations, and you do: two, three . . . seventy-two . . . That repeated movement now gives you something even more precious. It gives you a sense of time.

Imagine now that you have no memory. You see the spot in all its positions during each traverse, but you are trapped in the present and therefore have no way of knowing that the pattern at any moment has not always been. A sense of time demands memory as well as movement.

Time is a slippery subject. In the equations of physicists, and as proven by their accelerator experiments, time can be speeded up or slowed down by merely shifting one's gravitational field or velocity. Yet for daily life such concepts are hopelessly abstract. Time for most of us is a background, the ever-present beat of clock time that ticks off our hours, our years, our lives.

So what is time? What is this time that we actually sense? Is there any way to move beyond the otherworldliness of the physicist's notion of time and formulate a definition that is true to the richness of perceived time? I have found no better depiction than our imaginary voyage into the realm of spot and sphere: time is a binary of movement & memory. Here the "&" of course connotes not addition, rather the deep relation of synergy.

The equation that time is movement and memory is not a scientific proposition. It is a personal heuristic—or better, an inner song. It has helped me practice perceiving, and therefore exploring, patterns in time.

Many years ago, at about the same period of life in which the sphere bumped against my cranium, the equation for time blew into my neurons. I had been drawing seeds and flower parts, spider egg cases and webs, leaves and social systems. This last was not, of course, from my eye, but rather from what I could envision. Unlike my current devotion to reading, back then I was driven only to observe and record, both what I could see and what I already knew. I was searching not for new knowledge, but for new ways of organizing the knowledge I already possessed.

One day I was strolling dreamily under a blue sky filled with billowy clouds. The clouds were extremely low and racing to the east, their surging forms almost flapping in the wind. Processes, not things, coursed above. An epiphany knocked me to the ground. I lay there looking up, trembling with the knowledge of how mistaken I had been. By only sketching things in space, I had been neglecting time. I saw, or rather felt throughout my viscera, that I could and would have to redirect my search for general forms to encompass the realm of change, which creates with memory the shapes of movement, the metapatterns of time.

Time's metapatterns seem more complex than those of space because they can never be observed in a single ocular frame. They must be made by connecting several spaces. One of time's synonyms—the fourth dimension—suggests that this complexity derives from a higher dimension than the ordinary three of space. The dimensions can be raised as a holarchy of clonons: points build lines, lines build areas, areas build volumes. And what do volumes build? This is the classic impasse for imagination.

At minimum the dimensions can be constructed as a series of binaries: two points bridged makes a line, two lines bridged makes an area, two areas bridged makes a volume. But separating the volumes and bridging them just seems to make another volume. A physical, space-based answer of "other volumes" is worse than no answer; it halts the sequence by failing to create a unique new level. But a way to maneuver around the impasse is by taking the bridge between separated volumes not across space but across time.

To communicate about patterns in time, we commonly down-dimension time into space. Time is frozen in the horizontal axis of a graph and its pattern is read in a wiggly line that bridges space (y) and time (x). Time is represented by a clock whose motion we can read against the background of numbers. Time is implied by a linear series of photographs of a horse galloping in fragmented moments crystallized from a continuum. Time gets mapped into space.

Such mapping transforms the mysterious substance of time into more familiar objects in space that we can see, hold, and share. Psychologists call this the spatialization of time.

One universal map of time is a circle upon whose space can be placed the present, as a line or dot and either physically or as a point of attention. This map—the spot and sphere in our thought experiment—can be found in the circular astrological charts, in the circular ancient Chinese calendar, in the circular Aztec Epoch Stone, in the circular chronometer of the Age of Discovery, in the circular fifteen-dollar kitchen clock, in the circular disks on wrists, in the circular sundials. The apparent movement of the sun and moon and planets and stars probably embedded the circular motif of time in human consciousness long before the calendrical artifacts to chart these movements were invented. Now the calendar, in all its manifestations, has become a rival to DNA as one of the exquisite inventions of cosmic evolution.

Because of the details that must go on calendars, including perhaps spaces to jot each day's activities, and because of the convenience of printing and displaying a book of pages, calendars are most commonly rectangles. Large rectangles denote the months; smaller ones, the days stacked in rows and columns. This culturally created pattern resembles other cultural images rooted more in space—for example, the ordering of atomic elements in rows and columns of the Periodic Table, bricks in a wall.

Perhaps nowhere so clearly as in calendars can we know an image of time as space. In the calendar, too, one can find a wealth of time patterns. As we have dissected space patterns into their holons and clonons, so we may do for calendars, and thus for time.

Holons and Clonons in Cycles

If one had to choose some primordial time-atom that is at once physical, biological, psychological, and deeply social, it surely would be the diurnal day. The pulse of light and dark is the elemental brick out of which we construct the year, the month, and the week.

In the week, days are relatively few. European names hark back to ancient families of gods and the several bright dots wandering through the celestial zodiac. Monday, *lundi* in French, the day of the moon. Now, this and the other six words each call up powerful feelings (ugh! or ahhh!) because through many years we have learned to associate each day with particular activities and with proximity to other distinct days in the weekly cycle. Although mathematically clonons, psychologically the week's days are holons.

Sociologist Eviatar Zerubavel has traced the history of the seven-day week—the "prime invention of humanity"—from a world in which many alternative week lengths were and still are used. Humans have invented and lived successfully with anywhere from two to twenty days bundled into "weekly" units; the week's presence is what matters, not its length. Variations range from the New Guinea three-day market cycle, to the ten-day decades and hsüns (respectively) of ancient Greece and China, to the thirteen-day divinatory cycle of the Mayan sky gods. We are flexible.

The grand prize for time invention would go to the traditional Indonesian system of multiple weeks. The shortest is the binary *duwi-wara*, with alternating days of M'ga and P'pat. This and eight longer versions, from three to ten days, all spin simultaneously. Each variety of Indonesian week has its own sequence of named days, each day in turn contributing as a holon with a unique set of personalities, opportunities, and ritual injunctions. Each lived day, therefore, is a point in the concurrent rounds of nine cycles, a kind of alphabetic holarchy in time that creates the overall ritual cycle of 210 days.

Many African tribes developed market cycles of three to ten days, with four days being the most popular. This regularity rotated the market day around a circuit of neighboring villages. The difficulties of food storage and economies of sequential harvests may have imposed the need for a market rhythm with a small number of days. Overall, pragmatic social needs demanded a cultural cycle intermediate between the ones physically provided by the earth's rotation and the moon's revolution.

The moon set a template for our menses and our months. Although all English months except February are slightly longer than the lunar cycle of 29.53 days, historical adjustments in the lengths of the months made it possible to match their annual number to the closest integer number of moons out of the 12.37 moons precisely available. Because no month (except February, and not invariably) is an exact number of weeks, the day has become both the elemental unit within the week and, separately, within the month. And in the context of the larger bundle of the month, the day loses its holon character of name and personality, and becomes more of a clonon. Day twenty-three is just another brick in the month of time. As people function more as holons at home and clonons at work, so the day is similarly dual. Its character depends on the number of daymates within any given layer of time's holarchy.

Are months clonons or holons? From their number of twelve in the next level of the holarchy, we might expect months to act as holons. But sometimes months appear as clonons, just numbers on letterheads and checks. (I, however, prefer to name months in my computer records, even as three-letter abbreviations.)

In their very names months breathe as individuals. March, April, May, and June are the four names with the most ancient roots in our calendar's heritage. Even in the months whose names evoke Roman numbers (September to December), it is the splendor of the seasons and other memorable and recurrent themes of nature and culture that shine through, not their numbers (which happen to be offset by two—the twelfth month, December, is literally the tenth). Psychologically, months are predominantly holons within the year.

Unlike the month, the next larger layer is clearly a clonon most of the time. We of European heritage number, not name, our years. We routinely count and pile them up, stabilized by memory, like products

Portraits of time. A detail from Brueghel's *Triumph of Time* shows the months as the zodiacal holons. As time's cart rolls, life becomes death—leaves ahead, bare branches behind. Behind and beneath Brueghel's rendering are examples of time conceived by the Native American Dakota people. The annual cycles, either suspended from a line or chained in a line, here have a clonon character of sameness.

from an assembly line stored as the inventory of our lives. When we as societies attach a holon quality to years, it is not as individual solar circuits, but rather as lumps of decades. They are still usually called numbers—unembellished, as in "the sixties" (hey, who could characterize that?) and which actually applies to the first few years of the seventies too, or they are blatantly characterized, as in the Roaring Twenties. Lumping in tens brings them to life as collectively lived personalities, grand superpersonal zeitgeists.

When we are young, each year is a unique holon in the holarchy of self. But as we age, the years become too many to be individual actors in our story. We begin lumping them, most commonly into decades—one's personal twenties, thirties, forties—creating holons in this psychologically constructed level of time intermediate between life and its bricks of years. Those among us whose lives have been far from settled or full of progeny easily find personal ways to calibrate the years—but, even then, the occasional year slips by as a clonon, a gap between the big events, eventually a big blank. Perhaps society's recent tendency to merge years into decades for its holons of colloquial psychohistory relates to the convenience of doing the same for personal decades, because we only get to live for a handful.

Grouping years into tremendous units can also create a level of holons out of the years as clonons—again, signaled by the switch from numbers to names. For example, four Yugas form the Hindu cosmological cycle of 12,000 divine years, in which each divine year is 360 human years. In the countdown of four—Krita, Treta, Dvapara, Kali—one can recognize numbers with Indo-European roots. The Aztecs also had a creation cycle with a low number of major periods. Today, geologists recognize four largest geological time units within the whole of earth history, the eons: Hadean, Archean, Proterozoic, and Phanerozoic. This last, of 600 million years, contains three eras: Paleozoic, Mesozoic, and our current era, the Cenozoic, which began sixty-six million years ago with the impact that took out a lot of biodiversity along with the dinosaurs.

This look outward from the day into the largest units of creation has revealed both holons and clonons in the units at various levels, with degrees of clarity and ambiguity. As the second half of the film *Powers of Ten* did to space, we can in our minds continue the journey in time by descending into the details of time's units. What about the divisions within the day?

Originally the hours expanded and contracted, like accordion pleats, to fit a fixed number within the seasonally changing lengths of daylight and night. In medieval Christian monasteries the hours parsed the day into precise prayers and duties, and thus were very much holons. Even today hours can denote meaningful points or intervals in time, which may be shorter or longer than exactly one twenty-fourth of the daily round. What could be more lovely than to give personality to time by speaking of the witching hour, lunch hour, happy hour? On the other hand, hours, as official numbers with clock-locked spans, would seem to be clonons. They march inexorably during the workday, either watched anxiously or lost in the background. Hours, therefore, are ambiguous, at once like holons and like clonons.

The next smallest crisp unit gets lost as a clonon in a bunch of siblings, like tentacles of a sea anemone. Sixty are too many for minutes to emerge as distinctly functioning holons. (The television program *Sixty Minutes* gets split into a very few segments.) And from the minute, we have instituted layers of clonons thereafter all the way down, past the second, the microsecond, the picosecond, and as close to just this side of nothing as sophisticated measurements or imagination allows. After implanting the ancient Babylonian sexagesimal system of sixty for the first and second divisions of the hour (minutes and seconds), we then applied as a time pattern an iterated decimal division. This ad infinitum of clonons within clonons of the time and space measurement systems is as pure as a multilayered holarchy of clonons may get.

These holons and clonons found by telescoping out from the day and microscoping inward indicates the presence (in mind or reality) of analogous layers to those of space. And the similarity may be more than an analogy. We (or the universe) may have applied the great binary of sameness & difference to generate, respectively, clonons and holons in both space and time as a single, coherent process of unit creation. Additionally, because time is imaged and thus to a large extent understood as a projection into spatial patterns, it is no surprise to find a congruence between calendars and the general categories of patterns in space holarchies. Like the sea sponge regarded as a clonon by the population biologist but as a holon by the nesting fish, so too some of our calendar's units share both clonon and holon traits: the day set distinctively in the week or anonymously in the month.

The calendar can help us explore the similarities between time and space even further. For example, what are those lines between squares of days on a calendar? The breaks between days look like borders between lots in a suburban subdivision.

Taking Breaks

Compared to the physical manifestations of noon and midnight, dawn is as crisp as the edge of a cumulus cloud; the day jumps up with the sun. (If one wakes long after dawn, as Connie and I usually do, then night shifts into day at the sudden opening of the ocular port.) But when experienced fully awake, the glory of aurora only gradually unveils its enlightening gold, not slicing but evaporating the darkness. So, viewed closely, like the detailed ambiguity of a cloud's feathered edge, the transit from black through ruddy, into yellow, then blue sky is a fuzzy flow. In this way—crisp from afar, fuzzy up close—breaks in time are like borders in space.

A fuzziness less dependent on scale is the year. Physiologically, I would place the year's beginning in spring. With the bursting of first buds into leaf or flower, a chorus of copulating frogs, the first mild day of March, or the arrival of the first turkey vultures from south of the border, it is easy to label a crisp beginning to spring. But in point of fact the season of nature's general metabolic awakening is dispersed over a lengthening and strengthening of the sun that is stretched over time.

The earliest Romans and, later, the early Christians marked the round day's birth at dawn and that of the year during the spring. For Rome the original first month was Martius, our March. It initiated a block of four months of thirty days each—the entire calendar. Later, six months with numbers for names were appended. The linguistic legacy of months seven through ten remains: September through December. Still another calendar reform subsequently filled in the last blank zone with two additional months that now end in English with "-uary," which closed the counted circuit. Finally, in 153 B.C.E. the elected consuls, serving one-year terms, began the practice of taking office on the first of January. Thus started a new political year offset from and prior to the beginning of the former agrarian-based year.

A quarter year earlier still, around the autumn equinox, is another potential place to celebrate an annual calendar's transformation. Then the ancient pagan mystes were initiated into the mysteries of Demeter and her daughter Persephone, who had been abducted into the underworld by Hades. The fluctuating day of the Jewish New Year (Rosh Hashanah) falls around this time. Symmetric in their time holarchy, the Jews place a break in another cycle corresponding to the time when the power of light yields to that of darkness; each twenty-four hour day begins and ends at dusk.

It seems common for cultures to institute the breaks in their daily and annual cycles at congruent points within the smaller and larger cycles of oscillating fluxes of photons. For example, ancient agrarian cultures would begin the day at dawn and the year near the spring equinox. When the Romans enshrined two-faced Janus, the border god, as the initial month (following on the heels of the winter solstice), they also slid back the day's membrane from dawn to midnight. Today's International Calendar uses this symmetry between annual and daily points of furthest retreat of light for its official shifts in the numerical counts (only approximate for the year, of course).

By engaging such symmetries between the break points of day and year, cultures make these legal transitions seem natural; the patterns between two levels of time thus resonate and reinforce each other. But the fact that disparate cultures can define these breaks at different points in the continuous round of a cycle shows that these transitions are indeed arbitrary, like the political borders between many states.

The breaks in days and years are as finely defined as the legal lines between neighbors; today with atomic clocks we can cut these borders down to less than a microsecond. The crispness of the mathematical definition and the fuzziness of our psychological perception of the same border sometimes collide. The calendar may say the first day of spring, yet one region or another may get blasted by the biggest snowstorm of the winter. The printed grid now marches independent from nature. At least within the calendar—if not between the calendar and nature—breaks at different levels in the holarchy have been made to coincide, controlled by culture alone. In the International Calendar the beginning of the year also begins a month and a day. We can synchronize these cuts at all three levels, a task impossible for the early calendar-makers who counted moons as the intermediate level between day and year, unless they let the lengths of the official year float to match the moon.

Rivaling the visual crispness of the sun's fiery edge lifting above hill or sea is the reappearance of a crescent after the moon's monthly day

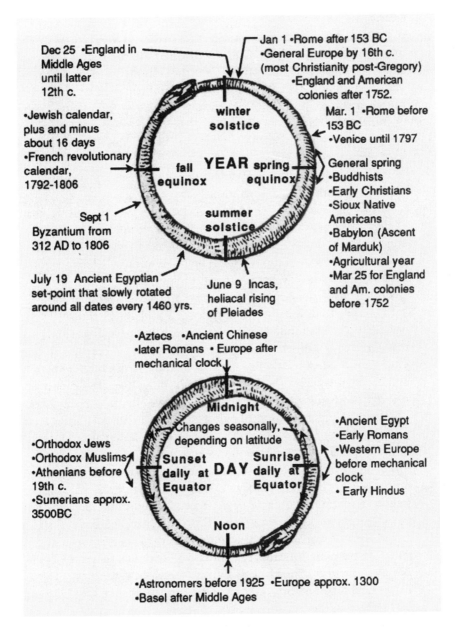

Beginning and ending years and days. Different peoples have differently placed the points where the snakes of time eat their own tails in the numerical and ritual breaks of years and days.

or two of cloistered retreat. Cultures that start the day at dusk also usually lock their calendars firmly to the moon. In ancient times, observers on hills high above town would confirm the first sighting of the silver sliver. From the Weda (a flower dance of the Nisenan tribe of North America), to the Ramadan (the Muslim fasting period), to the Jewish Rosh Hashanah, to Christian Easter, lunar-based festivals have permeated all peoples.

Reconciling the slips in correspondence between the astronomical year and moon's birth, the traditional Jewish calendar employs a nineteen-year superyear of the Metonic cycle. But for the early Roman crafters of what was to develop into the International Calendar, the complexity of maintaining coincident breaks in the sun's year and moon's month proved too tedious. The International Calendar lets the moon run its course. What phase is it tonight?

On my wall calendar the beginning of the week is Sunday, at the far left column. Traditionally, the Jews set the Sabbath (Saturday) at the end of the week, mimicking in microcosm the well-deserved break that God gave himself at the end of his labors of creation. When Christians needed to colonize some sacred time to call their own, they grabbed the day after the Jewish Sabbath. These two days eventually merged, thousands of years later, into what we all wait for: the weekend. With the globalization of culture, we might consider also incorporating Friday, the Islamic day of rest—deftly chosen at the foreside of the original sabbath—into the week's finale.

For most of us, psychologically, weekends terminate the week. A new week begins on Monday. As the seven-day circle spins, linked only to a repeating count of days—synchronized to neither year nor month—the weekends transform us from secular clock time into a sacred zone of worship, play, adventure, yardwork, or visiting relatives. They are times out of time. The short weekend punctuates the circle. No mere vertical membrane of thin ink between squares on a calendar, reducible to microseconds, the weekend is a thick skin of hide and fat between bodies of work weeks.

To pursue this analogy between breaks in time and borders in space, recall that borders have a variety of properties. One pair of properties has already been discussed: the distinction between fuzzy & crisp. Another is between casual & formal. A crisp cloud's edge is casual (nothing special). Our body's border, however, is formal (a precious organ, in this case). Now except for the somewhat flexible ritual of writing rent checks and the like, the beginnings of modern months are most analogous to the borders of crisp clouds: casual. The weekend, however—a thick, definite, and cherished zone of time—is more like the functional and formal borders of cells and animals.

Rituals at the beginnings of days serve as skins in time. Muslims pray, commuters slurp coffee. I have my bowl of oatmeal and a dose of *Nature* magazine. Quite formalized as a skin is the New Year's Eve inebriation and subsequent holiday (from holy-day)—ring out the old, ring in the new. For many of us the approximate ending of the year becomes a miniseason of travel and collapse, or seclusion, and always as much renewal as we can grab from the swift, formal shedding of the old skin of time. Near year's end the Japanese party hard with their culturally unique days of personally chosen celebrations called Forget-Last-Year and Forget-the-Coming-Year.

Matter and mind coalesce in forging the marks of the New Year, both on the calendar and in our skulls. Immediately on the first of January we start inscribing the new year number on everything needing a date, which sometimes seems to be everything we write. Furthermore, in the first weeks the new year really feels different; we are inside another block of time. Nonetheless, after a while, the piling of days distances us from the skin and the new tends to diffuse into the clonon quality of the preceding years.

We live with a similar pattern on a more personal level, based on the transforming skins of birthdays. Generally, the more special the break in time, the more like holons the zones on either side will be. When our years are few, each birthday forms a psyche-resetting mark, a transition that separates holons in our souls. But as our years take on more and more of the nature of clonons with age, then, correspondingly, birthdays lose their ceremony; they are no longer thick with import and remembrance.

Resetting the Great Count's Arrow

Although the birthday count consists of cycles, the count itself is not a cycle, unless considered in the context of life from dust-to-dust. Taken as an upward piling of years, shooting (faster and faster it seems) along the number line, the birthday count is a vector with a rate and a direction—an arrow. Along with cycles and breaks, arrows form the third and final metapattern of time exemplified in calendars.

The topic of arrows is vast and warrants an entire chapter, which it indeed will get. For now, even within the topic of calendars a diversity of arrows is found in time's holarchies. These holarchies are formed by cycles, of course, but also by break-terminated arrows. Arrows and cycles trade roles in the logical permutations of wholes and parts. Within the cycle of the day fall two arrows of the A.M. and P.M. counts: arrows within cycles. And before the Romans solved their computational problems by booting the moon from the calendar, they used to number days from high to low, in "countdowns" toward ritual lunar-locked days, such as Calends, the Nones, and the Ides; after each named ritual day (about four to ten days apart) the countdown would begin afresh.

Note that these countdowns were arrows consisting of daily cycles. Modern examples of down-counting arrows are a space shuttle launch, a rock-classic radio countdown, and children chanting the days 'til school's out for summer. Then the final permutation is arrows within arrows. For example, the history of architecture is sometimes envisioned as a great arrow with a sequence of smaller (and named) arrows of internally developing styles: the Romanesque, the Renaissance, the Baroque, the Rococo, the Industrial, the Modern, the Postmodern, and (I guess) the Organic to come.

All these ways of building time's holarchies will be explored more fully in the chapters that focus on the particular time patterns. For now, as an example of the kind of tunnel for the imagination into which a particular arrow can lead, consider the biggest arrow of our calendars. For most, death terminates the birthday count. The count may continue, however, for the noteworthy (1994 brought celebrations of the hundredth anniversary of Aldous Huxley's birth). The death of a very few exceptional personages can start another type of count. In Thailand the traditional culture increments the year from the death of Buddha—2,478 cycles ago as I write. In sections of India years are numbered from the death of a King Arthur figure of 2,073 years ago: Saka of Scythia. And 1,363 years ago the assassination of Yazdgard, the last Zoroastrian king of ancient Iran, set a clock that is still followed by the Parsees of Bombay. A timesmith might call these arrows that supersede individual lives, appropriately, the great counts.

The top of my wall calendar blasts the big arrow of the International Calendar in bold font, black on white: 1995. It is au courant to suffix this number with the letters C.E. for common era, which began at the end of B.C.E. (before C.E.). With slightest deference to the non-Christians of the world, C.E. and B.C.E. are intended to displace the long-standing A.D. and B.C., with their explicit references to Jesus as Lord (Domini) and Christ. In contrast to the Buddhist count, the C.E. count hangs around a birth, miraculous to the followers, surely numerically offset by some years, although how much is still being debated by scholars.

Deaths and births can seed great arrows—how about other types of events? For one, a perilous journey. In the Hegira 1,373 years ago, Mohammed, who had heard the voice of God but was not locally in turn heard, fled in disguise from hostile Mecca to a seat of power in Medina, where the ears were friendlier. Another great arrow began with creation itself. The traditional Jews calculate this great break at 5,756 years ago, plus or minus a few. Political revolutions can kick off great arrows, too. The exuberant leaders of the French revolution took a shot not only at resetting the Christian count but also at instituting a decimal time holarchy from months to seconds to match the pure efficiency of the metric system for space patterns.

Many great counts have come and gone on this planet. This current year would be 2771 if the ancient Greeks were still in charge (they initiated their count with the first Olympics). It would be 2754 if (dating from a circular furrow to bound the new city to which all roads would lead) Rome had not fallen. This year would be 984 in the fifth Aztec creation and 5108 in the Mayan Long Count—which is curiously only eleven years before the start of the current Hindu Yuga of Kali. These choices, now effectively extinct, show what happens to most great counts. The last Mayan Long Count we know was carved on a stela slab in a public square at Palenque approximately 900 years ago.

Will currently active great counts also go extinct? Are many to drop off the brink like so many languages, dress, and customs? In the globalization of culture and the competition for mind-space, one great count is reaching dominance. We may need coordinated time; we also need fairness. As one of many modern tragedies, cultures with deep memories, one by one, are essentially forced to disconnect their own great counts in the ever-expanding synchronization of the International Calendar. Even though all cultures in principle could keep their own counts for local rituals, the powerful and practical pressures from global economics and politics pushes everyone to match their calendars, like their clocks. By counting an arrow along the Common Era, we may be committing a common error.

When I met a Thai student and learned about the Buddhist calendar, the full implications of today's great count as a method of colonizing time struck me for the first time. From its deeply indigenous and exquisite culture, Thailand underwent a "cultural revolution" about fifty years ago. Along with requiring pants on male government workers and banning a satisfying, chewable native plant that turned teeth red, Thailand adopted the International Calendar. Their old calendar had "monks' days," linked to the moon, 4 days every moonth, and either 7 or 8 days apart (varying according to shifting fits to the moon's 29.53 day cycle). On these special days, the Thai people went to Buddhist temples for talk and prayers. Many still do, and the monks' days remain marked on calendars. For internal documents the Buddha's great count has not yet completely given way to the International Calendar.

We may eventually want to question the hegemony of other features of the (so-called) International Calendar: the structure of months, the holidays, the seven-day week, and the hours (the day and year seem pretty secure though). But the juggernaut that most transgresses diverse heritages is the Common Era count. It demands the most immediate attention.

The impact on lives from a great count comes about by its severing history into two parts, a before and an after. This simple binary can start playing parallels with others in our heads. If we number at all, this binary is inevitable. So to make any one personage from history— whether honored at birth, flight, enlightenment, or death—the pivot around which global time divides, tramples other worthies. The only way to be fair to all cultures is to not use any one of their great counts as the global great count.

Any great count tethers the time patterns held in collective memory. On a practical level, numbering years creates a filing system for the march of history's paperwork. But on a deeper level, memory not only records movement, it also interprets these records. The before and after of the great counts is a primary structure in our cultural memories. How easy to think of events from B.C.E. as having occurred in a magical, almost mythical time, a time of philosophers in robes, jackalheaded gods, and Racquel Welch running with dinosaurs, a time in which numbers for years even flow backward. No way better establishes a boundary between "home" and "foreign" territory in time than banishing all history prior to a great count forever behind a twilight zone of negative time.

What to do? We could restart the count.

Why should the Thai people have to lose the death of the Sitting One as the release of their history's arrow, replaced with dry news from the other side of the world? Would it not be a grand gesture—like seeing Earth from space for the first time—to reset the time with a date acceptable to all? The advantages of a brand new great count could be profound. It could have a powerful unifying effect on world psychology and project us into the future with optimism. But a problem arises—which year would we choose? Which could we all accept, or at least not begrudge?

Any event from very distant human history is out. Although many events led, in a sense, to universals—for example the origins of Buddhism and Christianity—all arose and matured locally, and their days of peak glory have waned. No, the event must be rather recent, probably from the twentieth century. Perhaps the launching of Sputnik or the landing on the moon? The first atomic bomb? Or the first time a signal was beamed from the Eiffel Tower (in 1912) and relayed around the world (which made it possible to synchronize clocks forever and thus electrically loop the noosphere)? I would personally vote for the original Woodstock—the rock festival in 1969—although I must admit it difficult to justify as any less parochial than Jesus' birth or Buddha's death.

Additional choices could even lie in the future. The impact of a comet would restart things, so to speak, but we might wait a million or more years for that. Certainly something wonderful or horrible will happen soon to the globe to warrant the year zero: just be patient. How about the year the carbon dioxide doubles in the atmosphere from its preindustrial amount? If itching for a restart, we could honor quantum mechanics and its inherent randomness: a lottery machine could roll out a random number of a near-future year. The world could watch on satellite television.

All candidates from past and future would likely stir vehement yeas and nays. Then what remains—the present?

There is one possibility, an event that would be the indisputable choice for an initiation, if it indeed could take place: we could agree to reset the count in the year we agree to reset the count. A new zero point would start when we reached an agreement (no small matter) that the International Calendar's current count be abandoned and a new one begun. Achieving consensus on such a globally charged issue would be the event celebrated and inscribed in memory by the count itself. The rising numbers would be an ascending crescendo to our ability to live together.

The issues of how, where, and why to set the patterns of calendars are similar in scope and impact to the issues that flow out of choices or happenstances in spatial organizations, which structure the world and our societies as clonons, holons, and borders. As we go global, what will happen to the great rituals, like Ramadan, Lent, and Hanukkah? Will they go the way of the ancient Greek celebration of the Panathenia? Will we be left bereft of special times, save for Superbowl Weekend? Can we initiate new Earth-scale festivals, perhaps around seasons already celebrated by many peoples—such as the winter solstice? Do we resurrect ancient magical celebrations upon the four cross-quarter points, with Halloween already a robust (if warped) start on one? Will the weekend go to three days? Could people work when they want and draw earnings by plugging into the system for varying times, like water drawn from a tap?

Calendars, with their cycles, breaks, and arrows, are time patterns humans have created, inspired by nature. The architects of calendars took their cues from the grossly astronomical. But cycles, breaks, and arrows appear, subtly or with panache, in more down-to-earth realms as well. These three metapatterns, in reverse order, will be the focus of the remaining chapters.

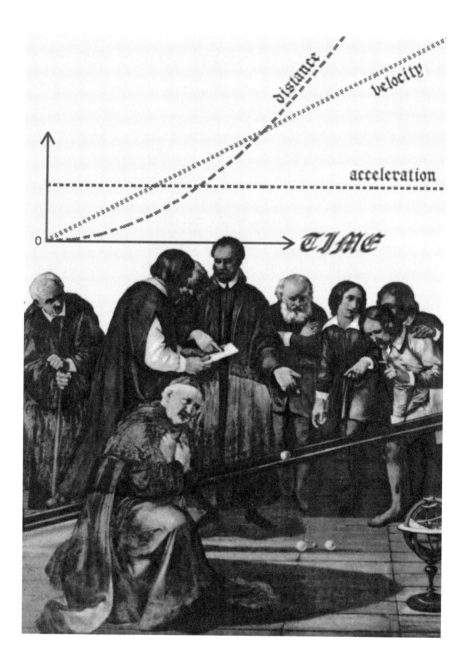

Galileo's experiment. Galileo used a sphere and a tube to discover the quantitative relationships of the three arrows that characterize movement through time: distance, velocity, acceleration. The man kneeling is marking time with his pulse. Galileo stands at center, with time coursing through his head.

Arrows

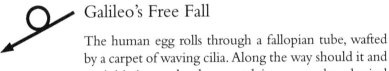

Galileo's Free Fall

The human egg rolls through a fallopian tube, wafted by a carpet of waving cilia. Along the way should it and a sperm merge—initiating a developmental journey—the physical trip nevertheless continues apace. This saga of a sphere traveling along a tube on its way to the womb can be read as an arrow of time.

Studies of traveling spheres have proved monumental to the development of physics. Ernest Rutherford beamed his "little cannonballs" (alpha particles) into the mysteries of atomic nuclei. Today hundreds of Ph.D.'s harnessed to humming machines that consume electricity by the power-plant-full splatter beams upon targets and create exotic particles that whiz away. How far we have come from the days when gaining fresh knowledge from the physical world required no more than a grooved board, a polished ball, a ruler, and, for timing, a wrist-pulse, improvised water meter, or steady singing voice.

In the honorable center of a commemorative painting is a little sphere suspended in midflight down a tilted wooden groove. The ball travels in a time pattern of a freely falling body. Behind, discoursing with students, stands the seer Galileo. How far does the ball roll after the first time interval, the second, the third, and so on? The thrust for such a question now seems commonplace. But it took wildly innovative, if simple, experiments to ignite the extraordinary idea of studying in refined numerical detail how a space variable changes in time.

Eggs rolling along fallopian tubes, and spheres along inclines—the real or imagined tubes by which things traverse physical space reveal a metapattern in time: the arrow. Time's arrows may vary from those relatively straight like Galileo's roll and the eight-minute path of light from Sun to Earth, to curved and often highly convoluted. For example, North America drifts away from Europe at several centimeters per year. Between them, in the North Atlantic, cold salty water at the surface plunges in underwater falls down to the dark benthos, thence creeping south, bifurcating and spreading, to eventually oxygenate the ocean deeps all around the globe. Above, at the molecular level, nitrogen, oxygen, and carbon dioxide molecules ricochet from collision to collision, with average flight intervals between collisions that can be computed in space and time. Inhaled, an oxygen molecule comes into the iron grip of hemoglobin and weaves through the blood vessels. Along a nerve axon an impulse of released ions glides. Upward in a tree, water is pulled along xylem tubes. Pheromones emitted from a female moth begin a lazy journey through the still night air. An ant scurries across the hot African sand, gathering food in heat no predator can handle. Salmon migrate up rivers and delicate monarch butterflies flutter hundreds of miles between north and south.

Arrows in physical space. A family of merganser ducks (with babes too young to fly) flees an intruder on a river in Alaska.
WATERCOLOR BY CONNIE BARLOW.

From the slow progress of ocean currents to the frenetic foraging of ants, the arrow of time is mapped in movements in space. Sometimes arrows ride along physical tubes (blood vessel, nerve axon, xylem). Elsewhere the tube occurs only in the arrow we draw with memory to indicate the path through free space. Thus things separated in space are linked via time, by arrows in visible and invisible tubes. The arrow is time's twin to the space pattern of the tube.

Cultures elevate feats of personal travel into mighty myths. The ancient Hebrews wandered the desert for the promised land. Jesus walked the crucifixion route. Mohammed fled from Mecca to Medina. The southwestern Native American tribes wove tales of migrations, perhaps a deep remembrance of humanity's spread into and across the continent. Buddhists make pilgrimages, Australian aborigines take walkabouts. Wars have key marches, football games crucial drives. Flocks of fans follow the journey of the Starship Enterprise with the enthusiasm of scholars following Dante's trek to the stars of the divine empyrean or Odysseus's peregrinations from Troy to Ithaca. The twentieth century has immortalized the peace walks of Mahatma Gandhi and Martin Luther King, Jr. Millions of Paleolithic enthusiasts (like me) have roamed with Ayla and Jondalar across the plains and up the Mother River of primeval Europe.

All such mythic journeys are recalled, on one level, as arrows in physical space. They connect distant places and thereby link the spheres of space into larger entities. But the myths also code for much more than the physical pulse. Journeys become metaphors for human accomplishments. The travel stories of the *Odyssey* and *Star Trek* endure because the physical adventure carries with it self discovery. The human spirit grew when Amelia Earhart flew.

These voyages are synonymous with mental arrows of learning and individuation. Thus arrows of consciousness are correlated with those that traverse physical space. Galileo watched the arrow of movement of sphere along tube. For humankind as well as Galileo the physical and graphed arrows became an arrow of learning.

Arrows in Property Space

Galileo's data of sphere rolling along tube can be plotted with time on the horizontal, independent axis and with distance on the vertical, dependent axis. This standard format yields the now-familiar parabola, the curved arrow of free fall. According to Galilean scholar Stillman Drake, Galileo's triumph was not that he studied change, but the change of change. And the change of change in distance—the graph's slope, or velocity—is also an arrow. Like distance, velocity increases with time during the fall. But instead of the curved parabola, the velocity arrow plotted as a function of time is a straight line that is inclined like Galileo's grooved board. Its slope (the change of change of change) is constant, which, when plotted is simply a horizontal line. Call it an arrow, too—the arrow of acceleration. Its time pattern equates to the statement that gravity did not change during the course of the roll.

Distance, velocity, and acceleration thus form a family of arrows. Only that of distance, however, can be seen in physical space. The others are arrows of change in "property space"—the space formed by the gradient of any measurement or calculation. Placing any property in a binary with time has become such a forceful tool of science that being without it is as unthinkable as a carpenter without a hammer.

A property space can be measured by many means: thermometer, barometer, meters for voltage, current, and resistance, or counts of populations, concentrations, and frequencies—all based on the tube of the number line. Just look at any issue of *Nature* magazine to see how we explore the unknown through the rising and falling arrows of the variables seen through the graphical lens of the space-time binary. With these roads, snakes, and lightning arcs of black lines upon white sheets we think about the forces and processes that mold the shapes of change, in everything from the electrical dynamics of the cell membrane to approval ratings of presidents.

After a day of executing computer runs, often with loosely directed walks through variables, searching for new phenomena, I'll bicycle home. Upon my arrival, Connie might start a line of inquiry: "How'd it go today?" When I answer, "Gettin' there," it means I really have no idea what or where "there" is. And I probably have been "gettin' there" for weeks. Often I never get "there" and have to back out from a boxed canyon of calculations and open another project.

Galileo, too, had to back out from some dead ends in that twilight zone of mazes wherein we seek for progress in science. For example, before discovering that the distances grew parabolically (in other words, with the square of the ratio between any two time intervals: 1, 4, 9, 16, 25 . . .), he tried a series of numbers based on a different assumption, with the ratios of 1, 5, 9, 13, 17 and so forth. We still have the page where he crossed out this guess. So, all in all, what is this arrow of understanding?

The path of knowledge is by no means as smooth as that of the rolling ball. The arrow of understanding is too complex to suffer the quantitative straitjacket of any sort of space–time graph. Yet an arrow

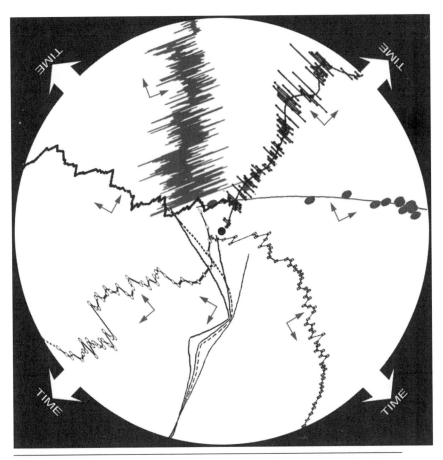

Arrows in property space. *Clockwise from one o'clock:* 4 months of methane emissions from rice paddies; 100,000 years of decrease in the size of human teeth; 60 days growth efficiency of wheat; the past 100,000,000 years of atmospheric CO_2 concentrations, modeled with 4 parameters; 8 years of a president's approval ratings; 1 day of the Dow-Jones stock average; 1 second of channel response in a neuron to a neurotransmitter.

it is. We can formulate its path by appraising the records of change (the memory of movement) using some misty sort of ranking. In contrast to distance, velocity, and acceleration (with their arrows of quantity), those of understanding and other properties like it—kindness, dedication—might be called arrows of quality.

One key word is development. Life probably offers the best examples. Look at the arrows of life—the maturing of a frog, the budding and blossoming of a flower. Frog and flower do happen to enlarge while developing. But surely any synchronous increase in size captures only a modicum of the complex and gangly arrows of development. Anyway, when "growing up" from the late tadpole to the earliest four-legged, tailless frog, the creature actually gets smaller—living off the energy stored in its tail while its mouth transits the functionless zone between round algae sucker and the wide grin of a tongued fly-catcher.

Plant physiologists distinguish a time-related binary: growth & development. Despite the often synonymous use of these words in common parlance, their distinction is a key technical point for agronomists. Growth is any change (increase or decrease) in measurables such as height, dry biomass, numbers of leaves, root depth, and grain diameter. The trickier concept of development deals with the phases of growth, its stages. Wheat, for example, has an early phase when leaves are initiated, which usually ends after eight to twelve leaves. A later phase, crucial to bread and pasta lovers, is the grain filling.

What makes these stages worth noting is that in the well-studied food crops their durations depend not on the ordinary time of calendar days. Rather, plants develop in accordance with thermal time (or photothermal time, which includes the effects of day length in addition to temperature). In thermal time (idealized), each phase of development gives way to the next when the plant accumulates a certain total degree-days above a base temperature. Hot weather speeds up the life cycle of wheat. Imagine the human life span so thermally dependent—we'd all scurry to Alaska.

Generally the binary of growth and development parallels the binary of quantity and quality. In music, volume grows, themes develop. As systems theorist Donnella Meadows and colleagues point out, humans ultimately need respect, joy, and love, rather than big cars and a closet full of clothes. They say, "a sustainable society would be less interested in growth than in development." This distinction may be a necessary condition for moving our societies along an arrow of melioration. We would continue to enhance the less measurable aspects of culture we cherish, such as art, music, literature, science, and architecture, while we concomitantly ease the insane drive to increase income, energy use, and power over nature.

Thinking of development as an arrow—or of any arrow of quality, including falling in love and bodily senescence—virtually calls up the metaphor of things traversing physical space. The mental time sense uses memory to place pictures in a sequence. Along a series of images as a tube in time moves the now.

Laurie runs for her health. I run computer programs. A CEO runs IBM. Getting ahead in life means picking up the ball and running with

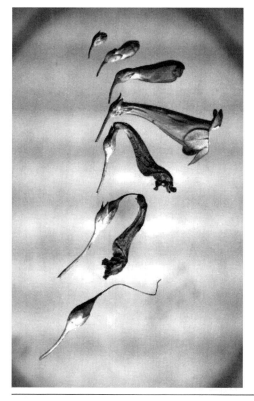

The biological arrow of development. Development of a penstemon, from bud through flower to seed.

it, maybe living life in the fast lane. Noses run, candidates run, and engines can run even when the car is parked.

Joseph Campbell advises, "follow your bliss," as if it were a road sign, rainbow, or special scent. One's bliss may be found by way of society's well-lighted roads of parental plans and progress within the hire-archy, on which one might drive for years, hardly questioning the map. But these can get crowded and constricted. One's bliss may lie within wild and rocky ravines, where brambles are common and not marked on the map.

Robert Frost wrote about roads taken and not taken; Emily Dickinson, about riding in a carriage with death. In Mary Catherine Bateson's concept of "composing a life," life should not be regarded as travel along a road with a set destination, nor success always measured by the directness of the course taken and proximity to a long-held goal. Rather, life is best lived as a meander, unexpected delights and new directions emerging from what might seem mere diversions.

Ever had your mind race at night? Had a train of thought interrupted? Had to stay the course? Bumped upon a problem, got stuck, and could not forge ahead? Was it finally smooth sailing? What is this line of reasoning? Is there a flow of argument? Welcome to the stream of consciousness.

Consider the course of a ripening grape. Arrows of quantity are manifest: sugar increases, starch decreases, seeds harden with protein, and chlorophyll dies while anthocyanins bloom, thus deepening light absorption. No one of these processes constitutes the ripening. Rather, the quantity arrows together make the quality arrow of ripening. We may measure any number of distinct changes, many perhaps indicative and useful in analyzing the ripening. Yet the overall arrow of ripening comes from the magic of language that constructs wholes from parts.

How many quantity arrows does one usually deal with when analyzing a quality arrow? A handful? Is that what it takes to make a good story?

 ## Linked Ups and Downs

Galileo's experiment of rolling sphere and grooved tube to discover the arrows of velocity and acceleration can now also be viewed as a study in the conservation of energy, the first law of thermodynamics. When a rock is dropped, its height above the ground decreases as its velocity increases. Technically, potential energy (proportional to height) is transmuted into kinetic energy (proportional to velocity squared). As the convention for a space-time plot makes the vertical axis of numerical values increase in the up direction, potential and kinetic energies are expressed graphically by counterflowing arrows of down and up. The success of free fall as a device for insight stems from the simplest possible holarchy of arrows: a binary.

Conserved properties of systems—such as mass, energy, momentum, charge—trade off among components via linked ups and downs. The gas level drops as the car's mileage goes up. A pool ball stops as the one it bumped ricochets away. At minimum the tradeoff is idealized as a binary, like a seesaw. But always a third flow, the increase of entropy, shunts some to spillage. As another energy example, recall that an airfoil lifts by separating the flow asymmetrically above and below. Above the wing the static pressure energy of air drops while its kinetic energy rises. Some of the fuel energy, however, passes into heat, as the airfoil must collide with the air to force the flow to separate.

The flower bud of a morning glory expands suddenly and by about ten times when it blooms. A complex biochemical dynamo directs the event. One substance produced by the developing pollen in the moist anthers diffuses down the anthers' stalks into the base of the yet tiny petals. There it promotes the production of ethylene, a growth inhibitor at this stage, which keeps the petals small. The anthers then dry as they mature. This slows their production of the ethylene promoter. Ethylene emissions plummet. Simultaneously, a growth enhancer (gibberellin) rises in the petals. The flower blooms. Thus a shifting balance between an inhibitor and an enhancer of growth controls the floral expansion.

The coordinated rises and falls, the synchronized accumulations and declines of control substances, account for growth, development, and just plain persistence of living things. A neuron fires, for instance, after it reaches a threshold of imbalance between excitations and inhibitions received from the vast network of other neurons it touches. Ups and downs in systems of regulation are not necessarily conserved as sums because the requisite mass and energy balances are usually traded elsewhere in the details of metabolism.

Causal loops of ups and downs are what is meant by the term feedback. If a change either up or down in one part of a system induces changes in other parts, which in turn recursively push the "cause" of the change even further in the same direction, then the feedback is termed positive. A positive feedback loop enhances a perturbation in either direction. By contrast, a negative feedback loop counters and therefore dampens the direction of perturbation.

The nuclear arms race between the U.S. and S.U. was a classic example of positive feedback. The direction of both arrows has reversed; they are now linked in the positive feedback of mutual dismantlement. My arms race occurs whenever I meet my lifelong arch enemy in ping pong, my brother. As the night wears on, the rounds intensify. Precious clumps of victories shift back and forth. Finally we

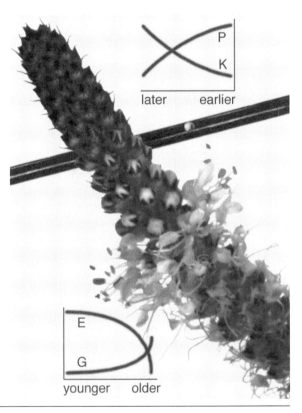

Counterflows. Physics links increasing kinetic energy (K) to decreasing potential energy (P). In blooming petals, falling ethylene (E), an inhibitor, allows the growth enhancer gibberellin (G) to rise—this coupling is not to conserve energy or mass but to guide the biological arrow of development.

are both playing at a peak escalation, at a dizzying level of slams and diving returns.

In climate, one example of positive feedback is the relation between ice and Earth's surface temperature. Say some agent—carbon dioxide—causes the surface to warm. Ice melts, exposing the underlying land or water, which better absorbs the incoming solar radiation than did the ice; Earth's temperature rises and so more ice melts, and so on. If values run away until all the ice is melted, the system is decidedly unstable. But convergence to a new stable value following perturbation seems to be the norm in recent earth history.

The scenario of rising Earth temperatures is, of course, of more than academic concern, with the burning of fossil fuels. A possible negative feedback in the Earth system is biological. All other things being equal, most plants tend to grow better at levels of carbon dioxide higher than that found in today's atmosphere. Plants in greenhouse

Positive feedbacks in climate. Ice and surface temperature constitute a positive feedback in the complex system of causal loops we call climate. An arrow of perturbation in one affects the other in a manner that loops back to amplify the original perturbation. This glacier in Alaska is an awesome arrow in physical space, but it is also an arrow in time: the age of the ice increases the further from its alpine source.　　　　　　　PHOTOGRAPH BY CONNIE BARLOW.

experiments thrive in elevated carbon dioxide because the more abundant this gas (their feedstock), the more easily it can be extracted from the air. And more vigorous plants could send more detritus to the soil. Such responses in removing carbon dioxide from the atmosphere could create larger carbon storages in living flora and soils, thus partially countering the rise. Were only the tale that simple, however. Concurrent higher temperatures with the elevated carbon dioxide would certainly increase the respiration of soil bacteria, which in turn consume soil carbon, sending that extra carbon and maybe some of the original right back into the atmosphere—a positive feedback.

Linked ups and downs in complex systems can be difficult to predict. In love, one can swirl into bliss as the beloved swirls too, wending into a binary of synchronized support and joy. But a lesson from the school of hard knocks is that ups and downs of love are not always so clear. That in truth you can be heading for a fall, just when you thought you were rocketing, makes up & down seem like two sides of an emotional Möbius strip. Shakespeare dramatized such a topology of two-in-one or one-in-two in *Macbeth*; the castle guard (Port) excuses himself for laxity in attending to the arriving Macduff's fateful knock:

PORT: Faith, sir, we were carousing till the second cock; and drink, sir, is a great provoker of three things.
MACDUFF: What three things does drink especially provoke?
PORT: Marry, sir, nose-painting, sleep, and urine. Lechery, sir, it provokes and unprovokes: it provokes the desire but takes away the performance. Therefore much drink may be said to be an equivocater with lechery: it makes him and it mars him; it sets him on and it takes him off; it persuades him and disheartens him; makes him stand to and not stand to; in conclusion, equivocates him in a sleep, and, giving him the lie, leaves him.

If the ups and downs associated with the arrow of drink cannot be unambiguously sorted out, how much more difficult the complexities of global climate change, the vagaries of politics and economics, the dynamo of love. And yet try we must.

Computers can help. They let us assemble intricate systems of interacting variables. Even a rather simple system to formulate—just beyond the binary—with three (or more) components twisted together in positive and negative nonlinear relations can produce the famously complicated behavior called deterministic chaos. Another approach uses huge numbers of units nearly identical in their behavioral rules. Complexity theorists study how nets of such units can create a mixture of properties, arrows up and down. They seek to understand the big arrows of stability and change as a holarchy of component arrows in so-called complex adaptive systems—language, culture, economics, organisms, ecosystems, and the biosphere.

One example of an amazingly complicated complex system that is nonetheless "adaptive" is seen in what is called the paradox of New York City. Consider: millions live with only a few days of food stocked in restaurants and markets and even homes. Yet every day enough replacement food is distributed and made available to make the whole thing work—all without any centralized control or master plan.

The human comfort with binaries and handfuls of holons means we often conceptualize complex systems by attending to a few graphs of data trends. From these we hope to glimpse how the whole operates. Much economic analysis that appears in newspapers is of this sort: interest rates are headed down, therefore . . . but stock markets are up, therefore . . . and if housing starts head down, then . . . but the national debt is up. Whither, therefore, consumer confidence?

I have a favorite shirt that is beginning to show its age. If I buy a new one to keep my wardrobe up, my bank account goes down. This lifts the economy by supporting the shirt manufacturers and retailers. But this also degrades the environment. The increased cotton production requires land with fertilizers and pesticides, and more pollution and energy consumption and carbon dioxide releases accompany manufacturing. (I think I'll keep wearing the frayed shirt.)

The same conflicts between up and down occur in issues larger than my shirt. Will demilitarization hurt the U.S. economy because jobs are lost, or help it by freeing funds for more productive feedback loops? Is clear-cut logging of old growth good because it is efficient and boosts the GNP, or bad because the boost is inherently temporary and biodiversity is lost? A thousand issues like these beset us.

We usually try to reduce complex systems into a few holons of related ups and downs. The ultimate model may be Galileo's free fall, with its tradeoffs between kinetic and potential energies—a true simplicity. All told, looking for linked ups and downs is a powerful tool for working with systems. It is an example of binary thinking that drives us into the beyond of new understanding. But up & down is not the only binary in the pattern of arrows. There is another pair, which, compared to the skinny lines of up and down on a space-time graph, packs real visual panache.

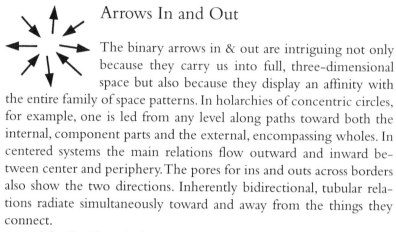

Arrows In and Out

The binary arrows in & out are intriguing not only because they carry us into full, three-dimensional space but also because they display an affinity with the entire family of space patterns. In holarchies of concentric circles, for example, one is led from any level along paths toward both the internal, component parts and the external, encompassing wholes. In centered systems the main relations flow outward and inward between center and periphery. The pores for ins and outs across borders also show the two directions. Inherently bidirectional, tubular relations radiate simultaneously toward and away from the things they connect.

And finally (almost), the progenitor of all the other space metapatterns also gives birth to the two arrows of in and out: spheres. Recall how they come into being. In the physics of atoms, stars, and balloons, two omnidirectional forces balance each other. These forces—various forms of contraction and expansion—map two sets of omnidirectional arrows.

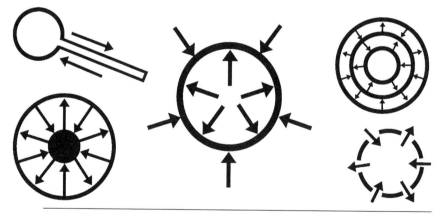

Arrows in & out and the metapatterns of space. *Center:* the balance of forces that make and sustain a sphere. *Clockwise from upper right:* two directions within a holarchy; two directions across borders; two directions between center and periphery; two directions along a tube going to and from a sphere.

More than all the other metapatterns, binary lives in the arrows of in and out. "In" and "out" are only tags for an extended family of processes: converge & diverge, focus & spread, gather & disperse, contract & expand, toward & away, accumulate & disperse, put together & take apart, shrink & swell, arrive & depart, to & from, unite & divide, merge & diverge, integrate & radiate. Many of these nuances of arrows in and out are not well suited for precision in the questions of science—at least not to the extent that up and down can quantify nonspatial properties. But in other ways their topological ties to space gives the arrows of in and out precedence over the numerical arrows of up and down. They set the stage for the universe we image and model. They may be more fundamental to thought.

Consider, for example, the engineering tool of control volume analysis. Its core equation for any conserved quantity reads: the internal rate of change equals the inflow rate minus the outflow rate. Understand this in all ramifications and you, too, could design power plants.

The two arrows of in and out occur naturally together. At day's end after ingesting bananas shipped from Ecuador, rice from California, and beer from Germany, my girth has expanded. The white king, whom Alice met through the looking glass, kept two messengers: one for coming, the other for going. A towhee bird calls for its mate to come and for rivals to go, to warn all around of the approaching fox and to get me out of bed at dawn to put bread on the porch rail.

Unless you print it, money obeys the law of conservation—checking accounts are control volumes. But manufacturing does not just conserve. It also creates. Absent inventory shifts, arriving steel is indeed conserved in the outflow. But the forms it came in disappear and toasters emerge. Whether or not conservation governs the sum of inflow, outflow, and internal change depends on what level you look within.

Green plants conserve the virtually indestructible element carbon in their systems of inflow, outflow, and growth, but not the substance carbon dioxide. Photosynthesis binds the hydrogen from water with carbon dioxide to yield organic carbon compounds, leaving oxygen to escape as a gas. (In simplified form, $6CO_2 + 6H_2O + \text{light} \rightarrow C_6H_{12}O_6 + 6O_2$.) The processes of cleaving and combining are essential to all metabolisms. Here we see the alphabetic holarchy deployed in time. Life scrambles components by using the arrows of taking apart and putting together, sometimes called catabolism and anabolism.

In the basic biogeochemical binary, the green plant takes in carbon dioxide and releases oxygen. But the plant also absorbs light, shunting it along molecules that converge to chlorophyll reaction "centers." Its roots pull water and nutrients toward the core stem. For outflow, the plant transpires water from stomates in leaves. Like the earth, green plants cool by emitting long-wave radiation, and like homes in winter, by convected heat swept away by wind. Finally,

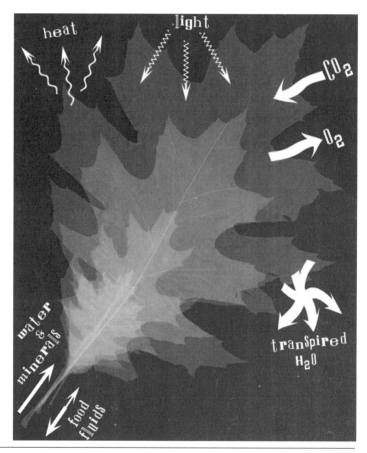

Arrows in & out during growth of oak leaves. Leaf samples were collected by author in the mountains of western North Carolina on May 6, 15, 20, and June 6. The earliest growth is driven by the mass and energy in phloem fluids stored in roots during the winter. The key to all net annual growth is the light-driven fixation of carbon dioxide into organic compounds. Transpired water is several hundred times the mass fluxes of carbon dioxide and oxygen.

at night, the flows of carbon dioxide and oxygen switch directions, as plants go about their business of using stored energy just as animals do. In many systems (a green plant is only one instance) a variety of arrows that radiate and concentrate coexist as vital components—holon arrows, perhaps.

Dynamics of inflow and outflow, of collection and dispersion, cast a general template for active systems. They can also become a matrix for imaginative leaps into swirling mind experiences. Watch circles spread away from where a raindrop dimples a pond's face and let go of worries as they too dissipate.

A practical side of mysticism with these arrows is the balance between contemplation and action. Jesus, Buddha, and Mohammed all took periods of seclusion in field, forest, or cave, followed by (or alternating with) teaching. The primal Egyptian pair of Isis and her brother-husband Osiris dramatizes the arrows of in and out through confinement followed by liberation. Osiris, entombed in a palace pillar, is freed by Isis, who conceives the hero-to-be Horus by her dead but at least debarked mate.

We map the arrows of expansion and contraction onto good and bad. Expanding one's consciousness is desirable, but being too "out there" can make one an outcast. At times healthy people withdraw, contracting inward for discovery and strength. Yet being withdrawn can turn pathological. The two directions create a certain tension in life, creative or otherwise.

The arrows of in and out have been invoked as a basic design for the inner principles of systems in time. In 1647, after successfully explaining celestial motions by means of centripetal (inward) and centrifugal (outward) forces, Isaac Newton thought one could go even further in applying the arrows of in and out as universals. He suspected that

> The rest of the phenomena of Nature . . . all may depend upon certain forces by which the particles of bodies, by some causes hitherto unknown, are either mutually impelled toward one another, and cohere in regular figures, or are repelled and recede from one another.

Goethe also envisioned the arrows of in and out as a set of universals. He sought to explain plant growth and form by alternate expanding and contracting processes. And he went even further:

> If God you want to find,
> first divide and then combine.

I interpret these words as pointing out (or into) an archetypal process of consciousness. We break things apart to study the holons and clonons within a holarchy and to figure out their dynamics. Then in our concepts we reassemble them. We come to know by way of a synergy of analysis of bits and synthesis of wholes. The model of arrows in and out depicts, in geometric images, the pair of deduction and induction. With the focus inward toward surrounded spheres or outward toward surrounding spheres, the pair reductionism and holism, too, fits this model.

Arrows in Possibility Space

The universe's most complex systems (ocean currents, bacteria, birds, soils, societies, Earth) are all what thermodynamicists call open systems. Such systems must exchange—by deluges of arrows in and out—at least energy and usually also matter with the environment. Why? Because of what many, following Arthur Eddington, have called *the* arrow of time—entropy, the cosmic disperser. The unremitting spillage to entropy must be countered within the system by importing fresh, available energy (which, in what seems a paradox, results in even more spillage to entropy—but that's life.) For those to whom the second law of thermodynamics stands supreme, entropy ultimately encompasses the other arrows the way the calendar's great count encompasses hours, weeks, and years.

The arrow of entropy is central because we live with it daily in the cooling of hot tea and the shattering of the cup when it is nudged off the table.

> Humpty-Dumpty sat on a wall,
> Humpty-Dumpty had a great fall.
> All the king's horses and all the king's men,
> Couldn't put Humpty-Dumpty together again.

Along with the thermodynamic arrow, physicists have listed other one-way flows in time. These range from the cosmological arrow of the expanding universe to the time-asymmetric decay of one particular oddball particle: the short-lived, neutral K-meson. Lists of *the* arrows usually range from a couple items to about a dozen. Is a handful of biggest arrows a sign of something fundamental about the universe? Does it reflect the (limited) state of our knowledge? Or is it another simple-minded instance of how we humans prefer to structure knowledge?

A potential rival to entropy's status as supreme arrow is one that usually makes everybody's list: the subjective arrow of time, our sense of time as a river flowing from past to future. Physicist Steven Hawking considers the subjective arrow of time as entrained in—and therefore subsidiary to—the thermodynamic arrow. In his view brains are thermodynamic devices hitched to the entropic fall of energy. Some psychologists disagree. To them, the human brain's ability to knit memories and internal images into sequences is not only vital but primary. Teachers, for example, exercise the cognitive abilities of children by having them correctly arrange pictures in a sequence—say, of an apple in various stages of being eaten.

Biologists, too, are engaged in debates about the overarching directions of time in their biggest story: evolution. Is evolution a spreading tree? Or is it like Stephen Jay Gould's image of a bush that has remained about the same size throughout most of time and that lacks, moreover, any apical tendency, any privileged lineage? He argues that although new branches emerge, the decimation of others just as luxurious results in no net growth of the whole. An opposing worldview

Teaching the arrows of time. Though my nephew Kevin could barely write his name, he pasted the apples in the correct order. That children are tested on this ability—and that they are rewarded with pumpkin-and-ghost stickers—shows the value we place on harmonizing our minds with the deepest arrows of time.

envisions evolution as heading somewhere, but then, where? Toward increased complexity? perhaps progress? Biologist John Tyler Bonner has provided some focus to what could be construed as evidence of increasing complexity by attempting to circumvent its qualitative aspects with some quantitative measures. He cites three large-scale, long-term, upward trends: organism size, the number of cell types in organisms, and biodiversity.

So how many arrows of time deserve recognition as the largest and the greatest? We might opt for one each from physics, biology, and psychology. Or a binary chosen from the universe as a whole might be nice. A big binary in the arrows of time is indeed the pattern that has cried out to many surveyors of cosmic evolution. Entropy is always one; the other most often has something to do with life and evolution.

To focus on life's ability to create bounded structures with relatively low internal entropy, Erwin Schrödinger invented the term *negentropy*. Buckminster Fuller proffered *syntropy* to stand for the geometry of concentrating, which puts it in opposition to the dispersing flow of entropy. Architect and futurist Paolo Soleri has diagrammed a creative evolution of "superstructuration" countering the "crushing megamachine of decay." To describe a life-related counterflow to entropy's inevitable drift downstream, neurobiologist William Calvin has coined the phrase "the river that flows uphill." Syntropy, superstructuration, an uphill flow—I am tempted to read a similar message into the Japanese national anthem:

> May thy peaceful reign last long!
> May it last for thousands of years,
> Until this tiny stone will grow into a massive rock
> And the moss will cover it all deep and thick.

Here we find not the collapse of order, like the one-way dissipative crash of Humpty-Dumpty. Rather, things build up, from small to large, from pebbles to rocks layered with moss. This upward process sings of stratified stability. Step by step, the development of complexity from level to level is an arrow in possibility space.

What can be said about the patterns set by this ratcheting up the steps of Bronowski's stratified stability? In my mind, such an arrow in possibility space has two essential components. The first is an arrow of contraction—a compaction of form. The second is an arrow of expansion—an emanation of relations.

Begin with atoms. Protons and neutrons compacted in nuclei form clumps of positive charge that accumulate surrounding electrons in distinct orbitals. The quantum topologies of these orbital shells breed possibilities at the next level for bridges between atoms, the expansion, in other words, to molecules. On the molecular level, atoms and small groups of atoms combine, break away, and recombine in the alphabetic holarchy. In Earth's earliest days this mixing of molecules that explored the possibility space of geochemistry soon hit upon the possibility of life.

The origin of life followed the building of complex, organic macromolecules from the stochastic shots into possibility space brewed in ocean foam, tide pool clays, or the sulfate oozes of a hot-house Earth. Knowledge about origins, however, grows in incredibly murky waters. We can say with conviction that several hundred million years following the estimated cessation of giant, sterilizing impacts, cells appeared in the fossil record. But how? Somehow the element carbon, with the help of hydrogen, assembled into chains short and long that adorned themselves with oxygen, nitrogen, phosphorus, sulfur, and others. Voila! Organic molecules were born. Some were lipids (useful for borders), others carbohydrates for energy and signals, proteins for enzymes and structures, and nucleic acids for codes. In ensemble they gave rise to prokaryotic cells.

With the gathering of molecules into cells, new relations emerged: reproduction and therefore population dynamics, food chains with mutual consumption—all the inflows and outflows of relational tubes to others and the environment—which in turn altered global and local biogeochemistry. It is at this level (if not earlier) that natural selection turned on, driving biological evolution further in possibility space.

Natural selection—the "Darwinian two-step"—is a prime example of the arrows in and out. Variation puts forth an expanded menu of possible ways of being. Selection restricts these, contracting the many possibilities into fewer probabilities and thence a remnant of actualities. (Gould's image of spreading branches and their decimation also contains the arrows of in and out.) Because life had to begin simple, it could only get more complex. Therefore, with life exploring various spaces of possibility in form and function and with natural selection weeding out certain trials, it is straightforward to see how an arrow of complexity could build.

The overall arrow from subatomic particles through prokaryotic cells thus made use several times over of this sequence: parts combine into wholes; these wholes engender new types of relations, giving rise to the possibility that the wholes become parts in turn for still larger wholes.

Several billion years after the birth of the prokaryotic cell, a major step in evolution tracked the sequence again. Thanks mostly to dedicated work by biologist Lynn Margulis, we now recognize that prokaryotes with diverse talents entered into permanent symbioses. These specialists—for photosynthesis, aerobic respiration, possibly mechanical motion—merged with relatively large host prokaryotes to form a variety of eukaryotic cells. (This step was surely a series of steps.)

The larger, more complex eukaryotic cell, with internal membranes and functional parts (organelles), radiated a new set of relations. These relations allowed, eventually, groups of cells with identical DNA to combine into multicellular organisms. Within these organisms, portions of DNA sequences were silenced and cells began to diverge in function, with the wholes undergoing life cycles that return to single-cell beginnings.

The explanation of why evolution "needed" the eukaryotic cell for life to complexify is unresolved. Was it, in particular, the advantages of a protected nucleus? What about the benefits of a host of organelles? The key point is that the primordial eukaryotic cell did have what it takes to exploit the possibility space of multicelled life. And eukaryotes then went on to direct bacterial evolution: The human manifestation of the eukaryotic type now engineers designer genes for patented bacteria.

The natural history of language is similar to that of life: words preceded sentences, but most words could only arise and evolve later in the context of a language with sophisticated sentences and stories. As the arrow of complexification makes its way, all the levels that were previously at the leading edge continue to explore possibility space too.

Untapped mysteries reside in the details within and relations between each level. Combining the model of spheres-and-tubes in holarchies with the model of arrows in and out may be a way to work our way into answers. Spheres (entities) with tubes (relations) form larger spheres. These larger spheres have their own distinct panoply of tubes—relations which did not emerge, which did not exist, before the lower-level spheres gathered.

With the emergence of organisms, arrows into possibility space could shoot for groups. Going beyond the clonons composing populations, ants and bees evolved colonies with castes and genetic centers. After several million years of brain growth and development, humans reached a similar, but unprecedented step on the ladder of complexity. Echoing in many ways, too, the merger of cells into organisms, the merger of humans into societies was accompanied by internal differentiation of forms and functions, but on the *cultural* level. This merger radiated another level of altogether new relations, such that social bands or villages could proceed along the arrow to another step of compaction into still larger units: states, nations, civilizations.

How many steps of contraction of form and expansion of relations moved our lineage from mere primate to whatever we may think of ourselves today? Any answer is fuzzy. That individuals belong to multiple groups and levels is one cause of haze in delineating the path to the present social holarchy. And social evolution continues at all previous levels, altering them far from their original states. Even less sure is speculation about what will happen.

Huge vistas of possibility space are yet to be discovered right here on Earth. The civilizations and cultures have only recently bumped into one another around the whole sphere. The arrows can test synergies between humans and computers, humans and nature, and humans together in kinds of groups scarcely imagined. Stay tuned. The steps, easy to see from hindsight, are tricky in foresight. Imagine looking at foxes, skunks, and even monkeys millions of years ago and envisioning that someday mammalian descendants would be sitting at desks in things called schools, learning to print the letters of alphabets and to arrange a sequence of cutouts from whole apple to core.

Arrows into the spheres of possibilities wend permanent paths only visible over the long haul and the big picture. We are probably too close to see our part. As Louis J. Halle mused:

> It may be that some future observer, viewing the history of civilization in a perspective that embraced the next million years, would see a consistent line of development from which the accidental circumstances of the close-up view had disappeared. Limited as we still are, however, to a close-up view that includes only five or six thousand years, we can hardly see more than the chaos of accidental circumstances that dominate the close-up view of anything, whether it is the constitution of matter or the evolution of life on earth.

The water that falls uphill.

Breaks

Time's Waterfalls

With a pop and a plop and a cosmic mind-burst of orgasmic light we are born. We switch on breathing, cough, and cry out into the gaseous, unconstrained, babble-filled maelstrom.

Jump-cut to a maple leaf in autumn. As the days shorten and nights cool, cells begin to callous near the base of the reddening leaf's stem. This programmed basal hardening yields two layers of crisp new border, which snap cleanly apart in a cold gust, or even from the leaf's own downward tug in still air.

Jump-cut to sixty-six million years ago. Burning with a trail of poisons through Earth's atmosphere, the comet explodes into the Yucatán Peninsula. A cloud of molten silicates tossed into orbit from a hole two hundred kilometers across condenses into skies of red-hot pellets of glass, which rain down over North America, turning air into oven, scorching the forests and searing life.

Jump-cut to a uranium-238 atom in rock ore. Stable for ten billion years since its assembly in some long-gone sun, the atom is about to undergo a change of life. Unpredictably (for the individual atom, but not for the ensemble average), an alpha particle blasts from its nucleus, which jitterbugs into its new shape as thorium-234.

Jump-cut to thinking. I am outlining an upcoming talk for a meeting of principal investigators growing crops in controlled environments. Doodling some titles for viewgraphs, I find myself launched abruptly into a daydream. Camping alone on a vision quest in the wilderness, I am attacked by a pack of wolves. (Is that how I see science?)

Jump-cut to the beginning of a song. The sweet tones start high. A flute and flutelike keys, in simple ancient harmonies, wend a leisurely descent along a path that promises, doesn't it, both the celestial and arcadian; then, after a concluding chord, a moment of silence tips and the prophet's voice begins the invocation: "Let me take you down, 'cause I'm going to [pause] Strawberry Fields."

Jump-cut to discussion and analysis. Birth, leaf abscission, mass extinction, radioactive decay, trance initiations, musical transitions—all these belong to the metapattern I will call breaks. Other names are possible: transformations, metamorphoses, punctuations, hinge points, phase transitions, passages, leaps. As usual, take your pick. I'll stick with breaks, a snappy word with Anglo-Saxon roots. It also instantly evokes a mnemonic image with a phrase—breaks in the arrows of time.

In their suddenness, breaks contrast with arrows. Together the two—break & arrow—make a powerful binary. The contrast may have structured the historical split of Chinese Buddhism (*ch'an*, or Zen) into two methods for enlightenment, the sudden (Southern School), and the gradual (Northern School). What is one of the hottest debates in evolution theory?—punctuated equilibrium versus gradualism. Thomas Kuhn's model for how science develops also fits this binary: ordinary times of "puzzle solving" occasionally ripped by paradigm shifts. T. S. Eliot mused about the world ending in a bang or a whimper. John Lennon sang about revolution and evolution—a widespread twoness in phrasing I have found in discussions of physics, forestry, and politics.

If time is a river, breaks are its waterfalls. Do waterfalls mystify, soothe, and induce trances because they are an archetype of transformation? We seek out these rarities and make them the destinations of hikes. There the water breaks from clear to white, from nearly silent to a roar; and the upper river dies in the plunge and flows reborn out of the chaos below.

Although the finger of metaphor can point to the apparent universality of sudden change—on diverse levels, such as rivers and consciousness—it can only scratch the phenomenon's surface. With reason one can probe further. One might infer that since the other meta-patterns had families of types, so should breaks.

A possible system for classifying breaks follows from the template of the figure & ground binary evident in so many examples. A comet punctures the sky and smashes into Earth; biodiversity is decimated. Or, day gives way to night, and wheat plants switch from carbon dioxide absorption to emission. Galileo signals for the sphere's release at the top of the sloped board; the sphere begins its roll. The clouds part, Wordsworth beholds the moon, and his mind is refreshed. In all these examples, a sudden shift in the environment causes or triggers a sudden shift in a system within it.

The change in the "ground" need not be abrupt to work a sudden change in the "figure." A gradual or sustained process (which can be imaged as an arrow of environmental change), may also cause or trigger a break in a system. Examples of systems with such threshold tolerances are many: the straw that broke the camel's back, divorce after years of irreconcilable antagonisms, the flight-or-fight response of threatened animals, the flames of revolution from gradually stoked social discontent, the slow erotic buildup to the release of orgasm, the onset of turbulence in boiling water heated from below, the formation of a leaf's abscission layer.

In a third manifestation of the breaks metapattern, a system breaks within a constant environment, because the system itself reaches a threshold that is internally determined. A star becomes a supernova; a radioactive atom releases an alpha particle; a leaf falls on a calm autumn day; I fall into a daydream.

Not all breaks, of course, can be classified in accordance with these three renditions of the figure & ground binary. Notably, there may be feedbacks between the relational arrows of entity and its environment; causation may not be one-way. Then too, it is difficult to map the structure of some breaks so simply. Take the standard computer routine called a "while" loop; a computation iterates until a selected output surpasses a numerical test, at which point the program boots onward to the next sequence of instructions.

It is worth exploring alternative ways of generating typologies for breaks. How do breaks appear, for instance, on space-time plots?

When the daytime photosynthesis of wheat switches to nighttime respiration, the carbon dioxide flux shifts from negative (going into the plant) to positive (going out from the plant). Graphically, this break is a crossing of the zero point on the y axis. But any relatively discontinuous change in a value of property space can be a graphical break. Most dramatic is the graph that resembles a waterfall—a (nearly) horizontal arrow turning vertically up or down, and then releveling to a new (nearly) horizontal arrow at a new value. Geophysical tension in fault rocks before and after earthquakes, reactions between molecules, and the transformation of egg into zygote can all be graphed as waterfalls.

A more subtle break is depicted graphically for the sun's altitude as it transits the sky. At noon the time derivative of the sun's position shifts sign: albeit very smoothly, rising becomes falling. Continuing the logic, mathematicians also note the changes in sign of a value's second time derivative—that is, the derivative of the first derivative—as a break in a system's behavior. The plenitude of possibilities for graphical breaks goes far beyond changes in values or signs of the derivatives of values. Chaos dynamicists, for one example, consider period doubling an important break, when the number of values visited during each cycle of an oscillating system switches from two to four, then to eight, and so on, eventually reaching the dramatic onset of chaos.

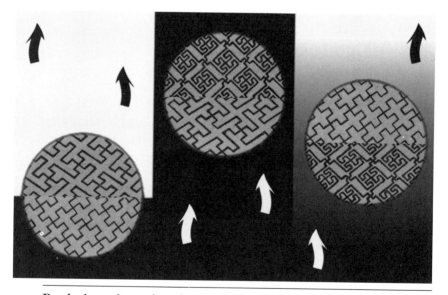

Breaks in entity and environment. A break in the surround can prompt a break in the entity; a break in an entity may be self-generated; or, a break in an entity may happen at a threshold level in a gradually changing situation.

Whether dramatic or subtle, in the end (and whatever the graph) is language not key? Language reveals how we regard the patterns of change. For example, our language makes a big deal of noon and midnight. Not only are these breaks marked by shifts between A.M. and P.M.; these are the only hours given names in addition to numbers. Language recognizes the transformation of caterpillar into chrysalis, thence butterfly, even though the organism (as species name) remains the same.

Language is better than graphs in expressing breaks in arrows that are more qualitative than quantitative. Birth, as a prototypic break, would be irksome to encompass with a space-time graph. A quantitative measure, such as distance from the womb, would hardly do justice to the rich holarchy of shifts at birth. Many other breaks involve changes essentially and primarily qualitative, and best identified through the action of naming.

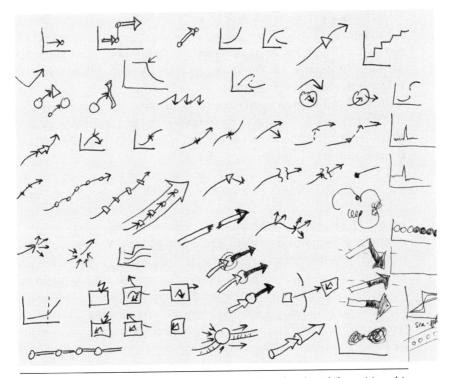

Visualizing breaks. Author's attempts to envision breaks while writing this chapter.

In fact, without breaks, spans of things in time (and ultimately, therefore, the things themselves) would be difficult to recognize. How easily we speak of the birth of rock-and-roll, the founding of the U.S. Forest Service, the origin of species. And language cuts through the fuzziness, making them real, distinct. Were there no real or perceived breaks in the arrows of relative continuity, the resulting mush of oneness would make speech, reason, and action as we know them impossible. By underpinning perceptions and concepts in the logic of naming, breaks in time act in ways similar to borders in space. Both are crucial for defining and knowing the things in space and their spans in time.

Time's Skins

Why should there be any connection between borders and breaks? After all, one is a space structure: we feel with our palms a tree's bark and observe cell membranes through a microscope. The other we assemble within the airy nothings of our minds: the eating of an apple from mere snapshots stitched in the memory of movement. Surely the patterns of time differ in fundamental ways from those of space.

In at least one category of cases, though, breaks and borders are virtually identical. The transit of something across a border is in fact always a break in time. Between states of temporary repose, first in kettle and then in cup, the tea is poured.

During a border crossing, both the crosser and the crossed experience breaks. Both the photon and the chlorophyll molecule transform when the first is absorbed and excites the second. For the crosser the break is quite significant, because the crosser invariably swaps its environment, going from outside to inside or vice versa. Disruption is less necessarily so for the crossed, the larger thing that loses or gains a little bit of something that it may need to acquire or shed in order to persist against the arrow of entropic dissolution. A snake gets a break from the pangs of hunger when it eats a polliwog, but the crossing is a doorway of death for the "wog." Walking through the door of a restaurant alters my mood dramatically, but for the waitress I am just one of many that day, more clonon than holon.

Beyond this most obvious identity can be found more subtle kinships between borders and breaks. The chapter on calendars introduced some of these. For one, closer and closer scrutiny—descending into detail—often evokes more and more fuzziness. Furthermore, although this rule may hold in a wealth of particular examples, individual cases vary widely in their fuzziness at the same resolution. Thus the things of both space and time can be placed along a scale of crisp to fuzzy: the crisp edges of rocks and the first menstruation, the fuzzy edges of hurricanes and menopause.

One scale-dependent spectrum from crisp to fuzzy is evident in the disputes about human origins. Because verbal classification itself demands a certain crispness (a fossil either is or is not in the genus *Homo*), our beginnings are naturally contentious. Arthur C. Clarke, in *2001: A Space Odyssey*, portrayed the moment of origin as the enlightenment of fractious apes by a huge, monolithic slab. But anthropologists are faced with a far fuzzier picture. The closer they look (meaning, the more numerous the finds), the less sharp the lines and the more vociferous the debates. One key moment in the film *Quest for Fire* showed a possible advance in being human with the beginning of frontal sex. This is my choice for the distinctive event that marks our origins—a most excellent theory because fossils cannot disprove it. The theory must, however, accommodate the fact that pygmy chimps, orangutans, and even gorillas also use the position . . . I'd better leave anthropology to the anthropologists.

Analogous to one theme of borders, breaks can function as bulwarks in time. If the Zen novice cannot pass the master's test, the novice can forget about the next stage.

NASA establishes review panels at various stages in the developmental flows of projects; these are decisive moments that can unfund dreams. The NASA reviews may, however, fund as well as unfund. Thus breaks, like borders, connect as well as separate. The action that spins a shroud, thread by thread, bridges the stages of caterpillar and cocoon. The snap of radiation reconfigures an unstable atom into a new element. Transformative rituals enable shamans and less adept connoisseurs of altered states to shuttle from the zones of daily reality to other realms.

A transformative ritual may be as simple as going out for juice in the middle of a work afternoon, or as formal as a !Kung circle dance to reach the healing trance. Like borders, breaks lie on a spectrum from casual to formal. For example, if it happens during sleep and no one notices, biological death is relatively casual, cell by cell. But look at the elaborate rituals that cultures emplace around, and particularly after, the moment of death. The two are as different as the skin of a snake and the edge of a cloud.

Mimicking what political borders do to those of nature, we often turn time's natural waterfalls from fuzzy into distinct, and from casual into formal. We dam the base of a stretch of rapids with a concrete catenary shell and regulate flow through a steel pore. Many fine minds from Sumer onward searched to synchronize the count-initiating breaks of the month and year derived from Moon and Sun. Out of the broadly spread biological and climatic events of spring we fashion a designated day, perhaps with villagers burning a straw effigy of winter and throwing the ashes into a river.

Puberty is another event (at least it used to be) on which human cultures stamped skins of ritual upon the breaks of nature. Traditionally, boys might exchange their foreskins for the rights to foreplay. Girls, too, in the ancient ways, underwent puberty ceremonies, on the average (with the horrific exception of pubescent clitoridectomies in some cultures) somewhat less dramatic than the boys' in terms of collective pomp that inflicts pain, fear, and all-out festivity. For instance, until recently the Jewish bas mitzvah, the girls' equivalent to the bah mitzvah of boys, was virtually nonexistent. Although the Jewish pattern may reflect the heavy male orientation of traditional Judaism, the widespread asymmetry may function to partially compensate for male biology presenting a fuzzier break than the crisp onset of menstruation. If the break is not obvious, we humans must contrive it.

I had no puberty ceremony. The Church tried to snare me into the rite of Confirmation—more like going for afternoon juice than a !Kung trance dance. But my eleven-year-old self wriggled away by declaring I had too much homework to do; my promise as a scholar would surely be ruined if I had to attend the weeks of preparatory indoctrinations. While still a young man I decided on my own initiation goal: I would count successes with the arrows of eros until I reached my age. But it never happened, because I got slowed down by love.

Fascinated by the formal and elaborate transformative rituals of cultures, I refuse to enter them myself. I make my own, at odd moments, usually fuzzier and more casual than the prescriptions of culture. For example, I did not participate in any of my three university graduations; instead of diplomas on the office wall, I have hung certificates from other moments. From my first ocean equator cross-ing on a research ship came proof that I was "duly initiated into the solemn mysteries of the ancient order of the deep." Another thumb-tacked paper certifies that I completed "the world's longest downhill bicycle cruise down the world's largest dormant volcano."

Evidence that we feel—subconsciously if not consciously—a deep relation between breaks and borders can be seen in an etching by that intrepid surveyor of transformation, William Blake. In his vision of death's doorway, Blake has visualized two states of time. The wall as a barrier of time separates life and death; a portal of transformation con-nects them. A juncture in time is thus represented by a boundary in space. Whereas at the plop of birth we literally pass through a pore, tra-versing death's door is (and only can be) a metaphor.

Native cultures of the American Southwest have developed stories of the time their ancestors reached the exit hole—the sipapu—to emerge from the underworld. The ancestors clamored for freedom,

Death's Door, *by William Blake.*

but the sipapu was too narrow for them to make the escape through their own efforts. Fortunately, sharp-clawed Badger dug it larger, and they crawled out and saw the sun for the first time.

Our myths of transformation are keyed to essentially the same picture: Plato's cave with its mouth of transformation between darkness and light, the shadow and the real; Persephone's furlough from the underworld and the onset of spring. The Bible narrates an ancestral couple's encounter with a ripe sphere and a slithering tube; the denouement sent them packing and out through the gate of paradise (literally, the around-wall). A recurrent theme in the age of emblem art is that of love liberating the soul from her cage. In the vision drawn by Rembrandt, which Goethe reproduced for the first edition of *Faust*, universal secrets appear as a mandala of sacred geometry within a window.

Vital to the journey of Joseph Campbell's "hero with a thousand faces" is the threshold, with its guardians and its dangers. The doorway guardians of temples symbolize this process in time. Thus in traversing from outside to inside, the worshiper mentally shifts from secular to sacred space. (Today the opposite direction, out into wilderness, may more faithfully rouse the sacred.) To enter the imaginary mental sanctum of Wonderland, Alice squeezed though a tiny door. In another adventure she entered through a mirror. In preindustrial Japan, an apprentice to a master craftsman would graduate with the ceremony of *noren-wake*, literally, "dividing the shop curtain." In China today, the slang for going into business is "plunging into the sea." Nearing the end of a rough time in life, one sees light at end of the tunnel.

Are border crossings in space the inevitable tool for portraying time's breaks in diagrams and myths? The doorway metaphor shows perhaps most revealingly the deep relation between the patterns of space and time. But more subtle expressions of their oneness abound. Consider the phrase, "I molded the clay into a pot." How the creative act of transformation is visualized resides in that mite of a word, into. Flying into a rage, falling in love—going *in* takes one from outside to inside. The structure is identical to the phrase "I poured the tea into a cup."

The skins of time we construct permeate everything we do. We make the week's end thick with a formal weekend. We ritualize the starts and stops of conversations. My Russian friends were amused when they first learned the habits of terminating face-to-face or phone conversations in America. "Have a good day"—who is this stranger to command so, or to care? "Talk with you soon"—oh yeah? "See ya"—that's pretty nondescript. More intimate goodbyes are embellished. We summarize, recall the pleasure of the dialogue's arrow, and reinforce the hope to continue the mutual journey at another date. And what about handshakes, hugs, and kisses?

Good jokes begin with hooks and end with punch lines. In the first session of a college class, students usually receive the course outline and an introductory lecture; in the last session falls the shadow of the final exam. Movies start with opening scenes; books start with prefaces (I put off writing my own until there was nothing left to write). Plays end with applause and curtain calls. To come to a satisfactory close, a jazz or rock concert must end with an encore.

As clothing speaks worlds about the person, so beginnings portend of the arrows to come. In the television sitcom *Roseanne*, when the family begins its slapstick high jinks around the dinner table as the theme song plays and credits scroll, I start laughing, get in the mood, remember their past capers, and anticipate the coming whole.

These skins in time play essential roles for the arrows they bound. The skins usually synopsize what is coming and what just came. Tones are set, summaries suggested.

Critical to science papers are the so-called abstracts. Like labels on cans, they announce what is inside. On his jar labels, Paul Newman not only tells us the ingredients, he explains how he invented the spaghetti sauce or salad dressing. In a science abstract a scientist explains what was done and what knowledge obtained.

Perhaps twice a year I put on a sports jacket and tie my ponytail tighter than usual. Connie invariably suggests my best black T-shirt for under the jacket. This is supposed to transform my interests into the foreign realm of administrative matters when I meet with those far above me in policy and power. Am I transformed? No. Do they think I am? Not a chance. Walking back to my office, I drift into a daydream. I see myself climbing the familiar slickrock of the arroyo that is my summer neighbor, or I fantasize a vision quest—me the seeker wrapped in hide of bear.

Animal skins donned in rituals have for millennia served as visual and tactile helpmates for reaching the desired transfiguration of the psyche. (Masks, too, fully concretize the oneness between breaks and borders.) A scholar of myth and religion, Mircea Eliade, has described the use of animal skins in various and widespread rites of initiation. For instance,

Holy transformations. The revelation of Moses is portrayed as God breaking through a hole from above. Central to the creation myths of the Native Americans of the Southwest is a hole from below—the sipapu—the site for transformation of protohumans into humans. The figure holding the staff has found the sipapu and is about to emerge. Finally, the tangible borders of masks create a change of state in the minds of wearer and viewer.

the word "berserk" originally signified a superhuman state of energy, and it derives from the Old Norse words for "bear's skin."

Social rules for moments of transition, such as graduations, weddings, and funerals, usually require the participants to be specially attired. Even I wore a jacket *and* a tie for my first holy communion at the age of five (and for my Ph.D. oral qualifying exam). The philosophy seems to be: to help imprint the mind with the sense of renewal or ending, transform the border. A glimpse into just how much these metapatterns of space and time, border & break, are linked in cultural concepts can be had among the Edo peoples in today's Nigeria. One of their principle celebrations, *Irhuen*, means "I put on a cloth."

Sequences of Stages

"Clearly, clearly the dye goes into my cloth." In the Niger delta the Edo sing as ritual aprons and robes are given and received, thus signaling a woman's maturity or a man's several transitions within the system of formalized age grades. Any culture must, at minimum, attend to two age groups—children and adults—and the transition that bridges them. Societies with more than two formalized types in the social lattices will provide institutions that ferry individuals across multiple moats, like marriages and political inaugurations, creating many types of cells in the body social.

Often I have watched the weaning of newly flying and still-awkward juvenile birds. It takes a towhee youth days to figure out that it shouldn't just stand there fluttering its wings and nagging while a parent pokes and prods for food in the duff only inches away. The rift from nest-bound dependence into dietary self-sufficiency divides the bird's life into several steps, with sexual maturity and its visual vestments, then courtship and domestic duty, coming much later.

Wheat development entails two broad stages: vegetative and reproductive. In the angiosperm version of puberty, wheat undergoes anthesis—the flowering that exposes and then mingles the ripened gametes. But at a finer level than this binary, plant physiologists have usefully distinguished a handful of substages, such as seedling development, tillering, and stem enlongation.

What is the metapattern here? Layers. Smaller arrows form larger arrows. In a holarchy of time's arrows, breaks both separate the component arrows from one another and connect them into wholes.

Breaks within a sequence may be a mixture of types, of course, from crisp to fuzzy, from casual to formal. The onset of egg incubation is crisper than is the end of parental feeding. And, as always, similar patterns do not necessarily mean that similar reasons underlie them. The overall arrow of a radioactive series—the steps from uranium to lead, for instance—is not a functioning whole with component sub-arrows. Radioactive decay simply *is*. In contrast, the life span arrows of wheat and people have evolved, in both biology and culture, to include a number of subarrows because this sort of arrangement is functional.

How rigid is the order of steps in a path? Digestion in humans ideally follows a one-way path. So do children through the grades of school. Atoms in a radioactive series, the march of the seasons, days of the week, and movements in a Beethoven symphony always follow the same exact path. Whether evolved in the cauldrons of biology and culture or dialed by the primordial hand of physics, the single arrow divided into a flow of units by breaks might be called the sequences metapattern.

If Walt Whitman were with me, I bet he'd sing the song of the sequences metapattern. He would see each detail of the prophase, metaphase, anaphase, and telophase sequence of a cell's arrow of mitosis. He would be there through all the explosive stages of a rocket. He would molt with the caterpillar through each instar of its growth and follow the path of aluminum from ore to foil. He would sing the stages of personal development in the theories of Piaget and Erickson, and the stages of cosmic creation in main-sequence stars. Embryo development from morula to blastula and gastrula; energy shuttled along the cytochrome chain in the membrane of a chloroplast; state formation, power consolidation, imperialization, and collapse in archeology; Carlos Castaneda's progression through the gates of dreaming; the sequential splitting of the four forces of physics in the early universe; the genetic leaps that altered the wild teosinte into maize; the primary, secondary, and tertiary steps that take proteins from amino acid chains into folded, globular forms ready for action; the computer bucket brigades of genetic algorithms; and for every human being the painful and glorious alchemical steps toward individuation—all these and more our friend Walt would extol.

In this list, the stages of rockets and the instars of a caterpillar come close to echoing the spatial pattern of clonons: the steps are repeated—although the rockets do get smaller and caterpillars larger as the sequences progress. Repetition (or near repetition) of events in a developmental sequence calls up the metapattern of cycles, to be treated in the next chapter. But in many examples the stages of a sequence are radically different, indeed so different that the stages

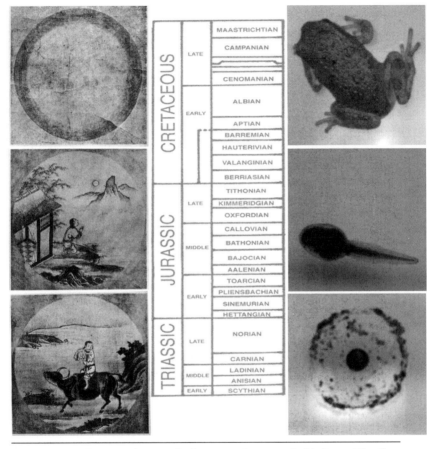

Sequences of stages in psychology, geology, and biology. The "ten oxherding pictures" interpret the Zen stages of enlightenment. Shown here are stages 6, 7, and 8: coming home on the ox's back; the ox forgotten, leaving the person alone; the ox and person both gone out of sight. The geologic time column (here the entire Mesozoic) is almost always depicted with time moving from the base to the top—thus paralleling the physical expression of time in the sedimentary rock record. The author further acquainted himself with stages by raising canyon tree frogs one summer.

could be considered as time's equivalents to holons in space. Space and time are thus even more directly intertwined. For example, the three largest time domains for an Edo man—manhood, Okpella-hood, and elderhood—are sequential for the individual and, since spread among many, simultaneous for the society; these time domains are holons in the social metabolism.

Many arrows consist of handfuls of stages. Like their counterparts in time—handfuls of holons—handfuls of stages in biology and culture can arise for express reasons. For one, when steps are added, complexity too may increase. Single-stage rockets gave rise to multi-stage ones. Substages may in some cases be added fruitfully anywhere along the sequence, in the middle or even at the beginning. For the more recent C4 type of photosynthesis (used by maize and sugarcane), extra metabolic apparatus was slotted in prior to the Calvin cycle of the C3 type (used by most trees). Kindergarten was added after the invention of the system of grades. It is tempting to relate the stable incorporation of further stages within an evolving arrow to the concept of stratified stability.

If the stages verge on the too numerous, thus creating potentially unstable complexity or other hazards, intermediate groupings or levels of stages may form. I do not know whether this has actually happened in biological evolution; there the designation of stages may say at least as much about our systems of thinking as about life itself. Different botanists look at the same wheat and peg it as having anywhere from two to twelve stages. But in arrows that we invent, grouping stages into a few largest ones is eminently functional. As an example, consider a NASA flight project with its mind-boggling multitude of tasks. Protocol includes an uppermost level consisting of a handful of phases, each lasting from one to four years: feasibility, preliminary analysis, definition, and design-and-development. Between these rigidly defined behavioral zones lie the yes-no filters of tough reviews.

Simplifying often helps us think. We can hold in memory the sub-arrows of an overall arrow, the events or processes, and then mentally juggle them for analysis and understanding—like we do for holons, the things of a space system. By networking these stages into a space-like system in our brains (and then handfuls of substages within each stage itself), we gain the same advantages of thinking in small numbers that drive our models toward the metapattern of low numbers of holons.

Were the five creation "suns" of the Aztecs—separated by cata-
clysms and rebirths of the world—a reasonable number to project as a
myth to remember? Did the parceling of Biblical creation into seven
days make it psychologically accessible, perhaps comforting? Consider
the divisions in the history of the English language into Old, Middle,
and Modern. The three stages are certainly arbitrary to a high degree.
And yet, the history was in fact broken by upheavals: the transition
from Old to Middle English more than eight hundred years ago is
seen in a wide-scale simplification, including the collapse of gender.
The mutation from Middle to Modern, around the time of Chaucer,
was marked by the "great vowel shift."

Why do symphonies most often come in four movements, with
three and five the next most popular choices? Why do therapy programs commonly feature twelve steps? Or pop songs a few verses? Or
books a handful of chapters? Or paragraphs a handful of sentences?

And what about our lives? The Dogon people of Africa view the
stream of a human life in stages, with breaks into adulthood at twenty,
elderhood at forty, and seniority at sixty. The track of university pro-
fessorship consists of subarrows called assistant, associate, and full.
Scientists in the current Russian systems (and in their Soviet predecessors) progress along five levels of rank.

Overall, from the mewling and puking infant to our infirm end
sans teeth sans taste, how many stages has life? English Renaissance
writers and painters took to the task of finding and enshrining an
answer. Those who found three followed the authority of Aristotle,
equating the stages of life with morning, midday, and evening. Four
had a natural analog in temperate zone cultures, in the progression of
the seasons from spring to winter. Five mirrored the number of acts in
Shakespearean drama; in this metaphor life was bounded by the
Prologue of birth and the Epilogue of death. The number seven was
ever fashionable, for the minds of the time were primed with holons
in sevens, such as the planets, the deadly sins, and the liberal arts.

A break in life does not necessarily lead to only a single subsequent
stage. A fork can lie in the road, a fork of choice. Minimally, such
points of decision have two branches. The history of morals evidences
a human tendency to posit a binary branching at the age of discretion,
a branching between lust for bodies or for books, between the sword
or the sphere, between virtue or depravity (not implying parallel binaries). Such a big branching was traditionally symbolized by the letter

Forks in life. Branches of trees spread into the possibilities of photosynthesis, those of evolution into form and function. The three phylogenies, based on molecular similarities, are drawn from diverse groups: bacteria, insects, and metazoans. The forks in human life—traditionally symbolized by a Y embellished with icons—spread into choices.

Y. And Y's on their sides appear on the modern space-time graphs crafted by theorists of dynamical systems. These Y's map points of change called bifurcations—but, this time, sans morality.

Food arrives at formal dinners in courses. The order is flexible between and even within cultures. Although the career of a honey bee usually begins with work strictly inside the hive and then shifts to foraging and finally to death, social exigencies that require more hive workers may prompt a shift in hormones that bathe the foragers in some potion that reverses the behavioral sequence. Our friend the fox may visit the compost pile at dawn one day, mid morning another, or (more usually) at dusk. The articles in *Scientific American* are intended to be read in any sequence. A bill in Congress may be sent back to committee.

The computer concept of hypertext uses trees of choices and multiple, simultaneous forks. Getting out of a linear sequence allows new structures for narration and for accessing information. The hypertext branching pattern may be the most effective way to navigate through cyberspace.

A similar pattern of many multiple branchings occurs in the flow networks of Earth's biogeochemical cycles. For instance, the reservoirs in the carbon cycle contain multiple entries and exits. In a food web, energy now in a beetle might next be ingested by a skunk or a towhee bird. Flows in complex systems of spheres and tubes engender this pattern. Board games exploit multiple branching to inject uncertainty and thus excitement into the contest—Chutes and Ladders, Monopoly's "go to jail" card. Finally, we think by way of multiple branching pathways.

Manifestations of branching patterns—moral bifurcations, the order of salad and the main course, the flows of carbon, the launching of daydreams—are all spin-offs from the prototypic sequence of a linear series of subarrows separated by breaks. Very generally, when choices lead to choices and to further choices, the proliferation results in treelike images. Perhaps the alchemical tradition's talk of growing through contemplation of the "tree of the philosophers" had to do with branchings of ideas and the self in possibility space.

Organisms develop by a sequence of cell divisions. In some cases, from any one division the two daughter cells give rise to lineages that become surprisingly different tissues or organs. The millimeter-long worm *C. elegans* is famous (at least to us scientists) because the fate of each cell born in each division, from the fertilized egg to the 959-celled adult, has been mapped. At the ninth Y of one lineage, for example, the daughters of one particular cell lead to body muscles and egg-laying muscles. Most of the worm's skin derives from one cell born out of a sixth division.

The surveyors of biological patterns also have discovered branching sequences in their maps of evolution. About a hundred years ago German biologist Ernst Haeckel drew (in lovely, curving lines) one of the first trees of the phylogeny of life. Today, the reticulate branching patterns of evolutionary relationships are derived as well as drawn by computer. Genealogical results based on protein data and DNA sequences from groups as diverse as magnetotactic bacteria, marine mammals, or the genus *Equus*, often look quite similar to the map of cell fates of *C. elegans*. This is because bifurcation is usually the rule. (Not exclusively or necessarily, of course. But for evolutionists of the cladist school of thought, binary is indeed the rule.) Species and higher taxa do not split like cells, but split they do.

Languages can also part ways. With deft comparative analysis, and at times using the same algorithms that compute the trees of evolution, linguists draw the myriad breaks and paths in the history of language. They have found how a few ancestral languages planted in North America, after thousands of years, branched into hundreds of others.

Both language and life can begin with one, and from that bring forth many independent lineages. The branching networks of both are at least partially generated by the summation of many small shifts, by mutations in DNA or in words, syntax, and meaning. Both exhibit extinctions: the giant ground sloth, Sumerian. Both can undergo massive and mysterious whole-system shifts: the Cambrian explosion of skeletonized, multicellular life between five and six hundred million years ago; the great vowel shift of English. Whether extinctions and whole-system shifts owe primarily to internal or external causes is still debated. But for some the evidence is clear: the current demise of regional languages owes to external forces, globalization; the collapse of life sixty-six million years ago was the deed of an errant asteroid or comet.

 ## Breakthroughs

In November 1907, while working in the Bern patent office, Albert Einstein was moved by what he later described as the "happiest thought" of his life: a person freely falling has no weight. He said that this idea "impelled" him toward a new theory of gravitation. For archeologist Denise Schmandt-Besserat, a 1970 report of tiny geometric objects inside a hollow bulb of clay from ancient Mesopotamia triggered the moment when "two pieces of the puzzle snapped together." The small clay cones, spheres, and tetrahedrons inside resembled the mysterious "trinkets" she had been finding in the dusty collections of museums—this clue launched her on an exciting path linking the origins of writing with these clay records of trade. And during Christmas 1981, Jim Lovelock, beset with criticisms of his Gaia hypothesis from evolutionary biologists, envisioned an answer. In the shape of a computer model he called Daisyworld—which demonstrated that planetary temperature could indeed be controlled by shifting populations of light and dark organisms (daisies served nicely)—Lovelock's solution came "fully formed, the way these things often do."

Such inspirations are what we in science all wait for, yearn for, and sweat for. Even from minor (but fortunately more frequent) epiphanies, we get ideas for new experiments to try, new angles to test. How to prompt such light-bulb moments is nearly impossible to teach, even to oneself.

Religion, perhaps, does somewhat better. That evolutionary capacity has had tens of thousands of years to cultivate cultural tools for helping humans achieve trances or lesser but still desirable fare. Like scientists, the spiritual explorers of states of consciousness seek to open the doors of perception and *see*.

How? By the drums and drugs of shamans, by all-night circle dances of the bushmen, by sensory deprivation and deep concentration of Buddhists or Christians meditating in desert caves. As the "ah-has" of scientists vary from the mild daily tremors of routine discovery to the rare earthquakes that shift paradigms, breakthroughs in the states of consciousness also span a spectrum. They vary from the frequent and ritualized altered states of shamans in healing ceremonies to the world-shaking insights from Moses' burning bush, Buddha's samadhi, and Mohammed's hearing the Koran.

Not only individuals but societies have been described as undergoing breakthroughs. The historian of art and architecture Kenneth Clark cheered those rare moments when, ignited by heroic collective energy, humankind would abruptly "leap forward," often in a single generation. He named one such time, that of eight hundred years ago, as Europe's "great thaw," from which blossomed the medieval cathedrals, scholarship, and trade.

Jean Houston, a modern pioneer in techniques to achieve altered states of consciousness, has called the medieval transformation one of the many "rhythms of awakening" that have pulsed through history. The metaphor—awakening—connects the biological arousal from sleep, with its implications for consciousness, to the larger, more nebulous blooms of history. Another metaphor for social awakening is birth, or rebirth, as in renaissance.

If such metaphors do not incontestably show us what is, they at least show us what we perceive. A rhythm of awakening prompted Wordsworth, speaking of the French Revolution, to proclaim "bliss it was in that dawn to be alive." (I felt that way too as a university student in the late sixties.) Similar feelings were expressed by the early boosters of what later would be called the Scientific Revolution:

> This is the Age wherein (me-thinks) Philosophy comes in with a Spring-tide. . . . Me-thinks, I feel how all the old Rubbish must be thrown away, and the rotten Buildings be overthrown, and carried away with so powerful an Inundation. These are the days that must lay a new Foundation of a more magnificent Philosophy.

The use of metaphors to draw attention to dramatic shifts in the mind of an individual or in the mind of society follows from the fact that breaks form a metapattern that spans levels. We employ the commonly observed, relatively simple, and therefore easily describable border crossings—dawn, birth, arising from sleep, inundations that sweep in the new—to point out the more interior, subjective, and complex breaks in the state of mind.

By way of breakthroughs we gain, we see more—new vistas, new worlds. Transformation becomes the goal at the end of an arrow of desire and effort. Psychotherapists search for insights, alchemists for transmutations of the soul, scientists for light-bulb moments, Zen students for kenshos and satoris. (The breakthroughs in Zen occur in a spectrum of frequent small ones, called kensho, and rarer large ones, called satori—a frequency-size relation also seen in earthquakes and scientific realizations.)

The underbelly of an insistent longing for transformation, however, can lead us mortals astray and provide food for the predators of power.

Human epiphanies. *Bottom:* an ancient Greek ritual of personal transformation in which the peplos robe was used. *Top:* the Encyclopedists portrayed
their collective enlightenment as an unveiling of truth; as Religion with *Bible*
slumps deposed, Truth stands, her unveiling assisted by Reason (crown) and
Philosophy (flame), with Memory and Ancient History looking on.

Thus political interests push newness as equal to goodness. Citizens were seduced by Reagan's "morning in America." In his wintertime inaugural Bill Clinton proclaimed that with his electors "we force the spring." It is de rigueur for those seeking or newly elected to office to evoke their coming as a breakthrough, enhanced by the yes-no switches of elections. Trotsky (and then Mao) extolled the concept of permanent revolution, perpetual transformation.

Business interests seductively push the new, too. Trained to be addicted to breaks, the masses hunger for the latest gadgets. I am tempted to trade in a serviceable, if battered, old car I already have, because here's the latest hit, the "renaissance of the American car," from that wonder of pattern-breakers—technological innovation (or is the difference just in the packaging?).

Delineating the current new age is a perpetual draw. Armies of slogans rather than troops are marched out to make good Trotsky's vision: this year is a new age, and next year is a new age, and on and on. Just since 1950 we have heard about the Third Wave, the Information Age, One-Dimensional Man, Megacorp, the Telematic Society, the Gene Age, the Control Revolution, the Space Age, the Age of Aquarius, and, as coming attractions, the Greenhouse Century, the New Millennium, and the Time-Compact Society of the anthill threshold. According to geologists, however, we live—and have been living (yawn) for the past two million years—in the Quaternary, which began at a glacial pace with the onset of the ice sheets. So relax. As Stephen Jay Gould reminds us, we always have been, are, and forever shall be in the Age of Bacteria.

One lives within multiple arrows of time, defined by their distinctive origins, as one lives in multiple space units defined by their borders.

Functionally, monumental events—the onset of the ice sheets, the founding of Rome, Buddha's death, Jesus' birth, Mohammed's flight, Woodstock—resonate through history and serve to organize social memory and therefore time itself. Immanuel Kant proclaimed Galileo's experiments with the rolling ball as the prototype of the modern scientific experiment. Today Galileo's experiments live on, not only as a prototype for instruction but as an archetype for inspiration. Memory swirls around key events, founding moments, originating kinks in the flow. Memory keeps the events alive and active and strong. Breakthroughs serve in this regard as centers in time, stabilizing the arrow that follows. Fragments of time focus around an origination like subsystems gathered around a spatial center.

Transformation of the psyche. For the European alchemists, lizards in flames served as symbols of renewal for their psyches. In Bali, girls enter trances and dance as goddesses for ritual gatherings.

Key events—riding a bicycle for the first time, discovering a fact of nature, falling in love—inform the successive arrows of human existence. For Jean Houston, a profound experience of childhood set her on a path of exploring practical means for entering (and exiting) altered states of consciousness. She had closed herself in a closet and implored the Virgin Mary to appear to her. Nothing happened, so she gave up. Then:

> Spent and unthinking, I sat down by the windowsill and looked out at the fig tree in the backyard. Sitting there drowsy and unfocused, I must in my innocence have done something right, for suddenly the key turned and the door to the universe opened. . . . My mind dropped all shutters. I was no longer just the little local "I," Jean Houston, age six, sitting on a windowsill in Brooklyn in the 1940s. I had awakened to a consciousness that spanned centuries and was on intimate terms with the universe. Everything mattered. Nothing was alien or irrelevant or distant. The farthest star was right next door and the deepest mystery was clearly seen. It seemed to me as if I knew everything.

Houston deplores the "lack of rituals of transition in our society." Through her work first with psychedelics and now with guided trance, dance, ritual, and dream, she has developed methods for people to "travel to their edges, there to fall off the known world and bring back news from the unknown." This heady quest for the rebirth of what she and others have come to call "sacred psychology" is the stuff of shamanism, discovery, and life itself. Her own life is an example of how a key experience can usher in a break that sets an arrow of commitment singing into the future.

Goals, like origins, can also function as centers for arrows of time in the present. The anticipated or hoped-for end of an arrow can orient its flow toward that point (or away from it, in the case of a projected disaster). This teleological point of convergence is key for the human arrows of consciousness. We set personal goals. Football teams drive to cross the end line. Science builds toward its "holy grails," such as finding the Higgs boson and elucidating the human genome.

Globally, we humans have never shared a goal. But today the vacancy of common purpose matters. As indicated by the rage to name the age, however, a sense of a massive social break happening now and now and now is rampant. We do seem to be in transit: the population explosion, for one, is undeniable. After a relatively stable and slowly advancing arrow for thousands of years, the population now is skyrocketing. It is clearly impossible for this trend to continue. Will it crash? If not, then it must level off. But at what level and when? And, more important, how? Whither the world and what can we do to nudge that whither for the better?

To halt the population explosion, a variety of breakthroughs have been proposed or prophesied. The Millennial Project offers seven easy steps for colonizing the galaxy. The Wildlands Project seeks to restore vast amounts of wilderness as biotic security, however the human bomb may tick. Social theorist Riane Eisler foresees a shift to a partnership model of society, ending thousands of years of a dominator model. Time theorist J. T. Fraser envisions a death control to supplement birth control. There is no doubt to me that we need radical goals. As the bumper sticker says, visualize peace. So let's visualize the end of population growth before it ends us. Then get really radical: visualize rivers clean enough to dip cups into and drink. Be optimistic about the break we are crossing. As Jean Houston says, "There's almost no myth I can think of—the search for the Grail, descent to the underworld, the hero or heroine's journey—that is as exciting as what we are living through right now."

Thank you, Jean.

Along with the biggest breakthroughs that occasionally blow minds—from society's awakenings to one's personal satoris—the smallest scales are also worthy of attention. Every day provides opportunities that can inspire us to focus on our own potential for breaks. Consider sleep and its phases of dreaming and then awakening. These commonplace shifts contain uncommon mystery. How do we wake up and continue where we left off?

You have probably already played with the optical illusion called the Necker cube. The stick figure of a cube can be seen in two orientations. The face first seen as closest can also be seen as the back face. Try the following exercise: First get comfortable with being able to see the cube in either orientation and then practice willing it to switch back and forth. Now—and this is key—try and maintain it in only one of its orientations. Can that single view be held? Or does it slip away? Most people can retain an orientation for only about three seconds before the cube spontaneously dissolves and reassembles in its alternate. This is an amazing phenomenon, a snap in the focus of consciousness from one state to another. For neuroscientist Ernst Pöppel it shows that "any content of consciousness has a survival time of only three seconds . . . and that within this span there is always only one content of consciousness." Furthermore: "Implicitly, then, the single content of consciousness can persist only a few seconds, before giving way, to be replaced by another."

Are the snaps of consciousness shown by the Necker cube a clue to the methods of Zen, which attempt to induce breakthroughs of understanding? Is solving a koan a sudden seeing of some of reality's other orientations? Does Zen take advantage of the fact that the mind is like the weather in Colorado? (If you don't like it, wait ten minutes—or in the mind's case, three seconds.) And what is the goal? Enlightenment? Then what?

> Before enlightenment, washing dishes.
> After enlightenment, washing dishes.

Is that a break—or an example of the next (and final) metapattern, a cycle?

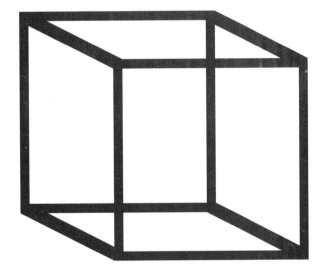

Necker cube and consciousness. Two choices for viewing are presented by the Necker cube. After you find them both, try willing them to shift back and forth rapidly. Then try holding any one viewpoint steady. You will fail—the cube spontaneously inverts into the other.

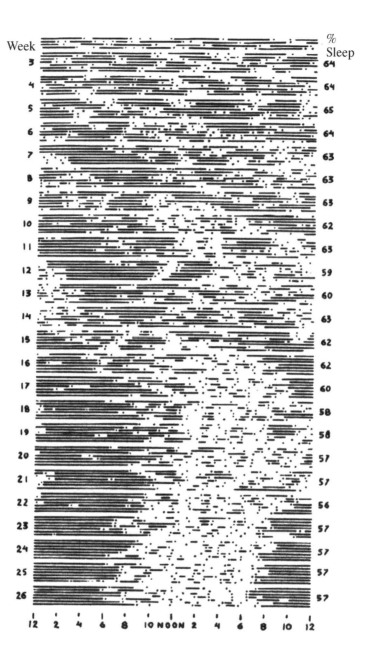

Sleep-wake patterns of an infant. Horizontal lines are sleep intervals, gaps are wakefulness, and dots are feedings. Although the percentage of time spent awake each day does not change much over the twenty-six weeks, its distribution clusters dramatically as the infant entrains to the sun.

Cycles

Rounds in Time

At birth, our lungs flood with air and begin their rhythmic duties. The heart, already fully active for many months, beats with new vigor. The suckling mouth, too, joins in the symphony of physiological cycles. With all—with the brain's gamma, beta, and alpha rhythms, with the mysterious growth cycles of stasis and spurt during the early years—we pulse: ethereally, grossly, statistically, chaotically, regularly.

The cycle of drowsiness and wakefulness in infants oscillates to a much more rapid drummer than provided by that mentor of daily cycles for adults, the sun. How, then, does this early, fast, and more sporadic flux of arousal and repose shift into the smooth diurnal cycle?

It takes about a half year for the baby's sleep-wake habits to coalesce and stabilize into the predominantly binary beat. The newborn learns to match its biological rhythm to that of its parents, responding especially to mother's heightened attention and jocularity during daytime feedings. Baby also entrains to the sun. What are initially short spells of wakefulness combine into larger chunks, which themselves tend to concentrate once a day. At first this zone of coalescing wakefulness is not fixed to an hour, but wheels around from day into night into day. This drift, which completes three to four rounds before stabilizing into the diurnal pattern, owes to our species' internal biological clock of 24.7 hours. Eventually, not able to lick the sun, the baby joins it.

From the sun beams a supreme archetype for both the first metapattern, the sphere, and its equivalent in time, the final metapattern, the cycle. And equal to Sun as an archetype for both spheres and cycles is Moon. How and why evolution entrained the average duration of the reproductive flows of women with the lunar round (to the tenth of a day) remains an awesome mystery. From the waxing and waning of its soft glow, between a shrouded absence and a bright beckoning disk, females and males together have reaped the deep mythological theme of the eternal return. Worldwide, rituals have helped harmonize mind with matter, humanity with cosmos, through congruence between myth and moon. We behold and embody from Sun and Moon perfect sphericity and primordial cyclicity.

When we shoot satellites into the formerly untouchable above, home of the heavenly cycles, and turn their electronic eyes back upon Earth, more cycles are revealed. Giant white pancakes sweep across seas and plains, gyres of clouds that trace the wind's spin around highs and lows of atmospheric pressure. The lowest of the lows form the centers of hurricanes, revolving in periods measured in hours. Much more slowly revolve the ocean's thermohaline circulations: after plunging from the surface during polar winters, deep waters ooze through the global benthos, then slowly rise and return, over, on average, a thousand years. Still more sluggish, taking millions of years to complete just one up & down cycle, convection cells churn the earth's deep regions.

The many sizes and speeds of cycles in space sets the head spinning. Galaxies, planets, electrons and other quantized particles have characteristic spins. Eyes rolling, maple seeds whirling; mating dances, blood circulation, a hand turning a screwdriver; corkscrewing bacterial tails, spinning algal *Volvox* spheres, whirring wasp's wings, twirling frisbees; clock hands, turbine blades, Wheels of Fortunes, laundry drums, microwave platters, Tibetan prayer wheels; cogs, gears, axles, shafts, electrical circuits, the blurs of the centrifuges of molecular biologists; computer disks, tape disks, the Dead Sea scrolls, rolls of fax paper, CDs and their obsolete vinyl ancestors; circulating chloroplasts in plant cells, chipmunks at play; Olympic track laps; tornados and kitchen blenders—all these and more old Walt surely would have extolled were he sitting with me now.

We take cars for spins; politicians work spin control. Buddhists call their doctrine the great wheel and portray it as a chariot wheel. One of the cosmic consequences of any spinning in space is the confining of matter in motion to a finite region. Cycles thus define and make things. Equally, things contain cycles. The word *cycle* has a common root with the word *circle*. Thus a round in time is understood as essentially synonymous to one in space. In the language of metapatterns, the cycle of time is sibling to the sphere of space.

The cycles within cell parts, cells, and organisms are vital to integrating the respective entities. Circulation circuits, such as the protoplasmic whirlpool of organelles in plant cells and the molecular rounds of chemical processing (notably the Calvin cycle of photosynthesis in chloroplasts and the Kreb's cycle of respiration in mitochondria), fix the organism in time as well as space. Flows of matter and energy connect, diverge, and reconnect. A swan's legs, for example, have what is called regenerative heat exchange. Vessels that move blood toward the feet are in close thermal contact with those that return blood, thus enabling the bird to strip away and recycle much of the heat from blood destined for an uninsulated part of the body.

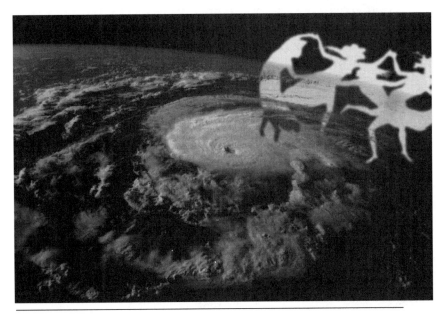

Hurricane dancers: spins in physical space.

In the earth-scale metabolism of Gaia, biogeochemical cycles shunt the life-supporting elements back and forth between reservoirs, in and out betwixt the geology, chemistry, and biology of the planet. The atmosphere's nitrogen, which is continually oxidized by nitrogen-fixing bacteria and lightning, would all dead-end as nitrate in the ocean were it not for the counterflow of nitrogen back into its pure gaseous state by other bacteria in soils and oceans. The near-perfect closure of the cycles of carbon, hydrogen, oxygen, nitrogen, phosphorus, and sulfur—with life at key nodal points—provides a base for the seemingly effortless persistence of the living biosphere and its evolutionary heritage. This "genius of recycling," as ecologists Allen and Hoekstra call it, "is a strategy both for escaping the constraints of scarcity and for regulation."

As life discovered long ago in the trial-and-error laboratory of evolution, cycles preserve the entity. Even in the trashy concrete canyons of the New York City ecosystem, bright blue bins are appearing for recycling newspapers, bottles, and cans. We are finally learning: what goes around, comes around.

Metabolic loops in space—within cells, organisms, cultures, and Gaia—carry not just matter and energy as so many nanograms or gigatons. They carry patterns. They carry circles of influence, circles of causation, feedbacks. Conversation is such a circle. You speak to a friend, she responds, and the circle empowers itself. Rainfall spurs plant growth; the water pulled from roots to leaves returns skyward to rain again and sustain the growth. In cells, genes encode for proteins that activate other proteins that may feed back and turn genes off or on. Engineering firms iterate between their designs and the design process, between plant performance and design philosophy.

In the menstrual cycle, diverse body regions (central nervous system, hypothalamus, pituitary, ovary, and uterus, as well as environment) form a loop in body space around which flows the substances of stimulation and inhibition, such as follicle stimulating hormone, luteinizing hormone, estrogen, and progesterone. These hormones rise and fall monthly, like flowers that open and close daily for their nocturnal or diurnal pollinators.

Such hormonal ebbs and flows can be charted on a space-time graph. The space here is "property space": the space of any and all values and measurements other than distance. Ups and downs in property space are a useful way of contemplating cycles. For example, basal body temperature fluctuates in the menstrual round; tracking its pattern is crucial for those who enjoy making love without contraception or children. By monitoring the relatively low basal temperature during the egg's maturation, then the body's slight cooling, followed by the dramatic rise in temperature after ovulation, many women can accurately determine where they are in the monthly cycle. Counting beyond the thermal marker of ovulation for a few days to the egg's death signals the beginning of a period of "free time."

Cycles in property space are crucial to the physical laps through which we put the fluids of industrial metabolisms. Consider, for example, the hallmark of the industrial revolution—the steam engine, liberator of the motive power of heat. The key cycle involves water and steam forced around a series of expansions and contractions, of pressurizing and depressurizing, of heating and cooling, of elevated and lowered entropy. The engineers who squeeze motion from fire use heartbeats called thermodynamic cycles, such as the Brayton cycle, the Rankine cycle, the Sterling cycle, the Otto cycle. On binary graphs these cycles can be studied and potentially improved (in some cases by

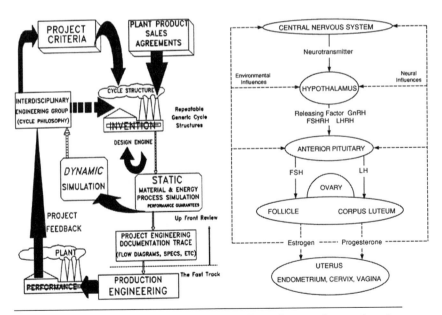

Cycles of causation and control. The discipline of power plant engineering uses a cycle of design, simulation, building, and monitoring. Hormone feedbacks in the menstrual cycle pass among the central nervous system and various glands.

strategically intertwining cold and hot pipes to save energy, like the vessels in a swan's legs). Thermodynamic cycles close the systems in time, giving them shape in property space, as essential as their cycles in physical space.

I look forward once a year to a special graph of temperature. Published in the *New York Times* in early January, this plot of New York

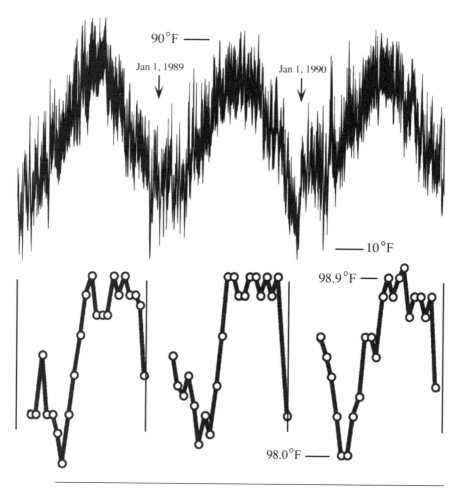

Two cycles in the property space of temperature. A graph of the daily lows and highs in New York City for three years. A woman's basal body temperature during three menstrual cycles. The vertical lines mark the start of her periods.

City's highs and lows for the past year shows the binary of sameness & difference. The seasonal cycle inexorably ebbs and flows from cool to warm to cool again, entrained by the sun. But "unseasonably" hot and cold spells are driven by the chaotic dynamics of regional climate, an infinitude of detail that never repeats. This pairing of repetition and variation is thus a manifestation of two different scales in time.

Perfect cyclicity, like perfect sphericity, resides in the mind. Yet nature does approach these ideals with the great rounds in time and space from moon to grape. We revel in difference, and yet, as often necessary, we filter it away to focus on sameness. One cannot body surf the same wave twice. But one can always run back out for another.

Cycles conjoin with arrows and breaks to form a triumvirate in the metapatterns of time. In the fantasy piece that introduced the chapter on calendars, only when the spinning orb was given a spot could memory merge with movement to form subjective time. The count and the interval, or the break and the arrow, alternate in a larger time pattern, the cycle.

Cycles are the heartbeats of understanding. A thing perceived is just an event. Repeated, it opens up to the instruments of science, the ruminations of philosophers, the imaginations of shamans.

Cycles Propelling Arrows

Contemplating cycles invites one to make the rounds of imagination. Like the sphere, the cycle can be a fount of inspiration. A sudden awareness of the omnipresence of cycles has at times stopped me in my tracks like a spider's web stops a fly, leaving me vibrating with a vision.

The sequence that sets the spark, however, may be quite ordinary. I remember a slice of New York pizza in a crowded, stand-up eatery. Munch slurp munch, munch slurp munch. One of those inane pop tunes that replicate in the head as mind viruses for unexpected (and usually unwelcome) replay was blaring—a blast from the past that should have been sunk twenty thousand leagues under the sea. Oh baby it's Saturday yeah yeah; oh baby it's Saturday yeah yeah. In the corner the dough juggler worked the ever-widening floppy disk up and down, up and down, while the oven master slid pizzas in and out, in and out. Customers entered and left, entered and left. Across the counter coursed (as vital to our economy as blood to our bodies) the most worshipped cycle of all: dollars in and out of pockets and cash registers. Ring ring ring. The web had hold of me. The surrounding events popped into a single presence that permeated the multilayers of everything, the parent metapattern from which the many siblings had sprung—indefinable, but precisely felt in my pizza parlor epiphany.

For some, such moments can trigger epilepsy. For me, they have fueled a passion for the mysteries of our planet's biogeochemical cycles. The carbon cycle, for instance, offers many unknowns to the would-be puzzle solver. It is so complicated that, out of the total we inject annually into the atmosphere by driving cars, building widgets, and playing with matches in forests, we cannot yet say where the one-to-two-billion tons of "missing carbon" has gone. Every year, we add more carbon dioxide to the atmosphere, but only some of that increment remains. Has more than we expect disappeared somewhere in the sea? Or is there some natural, as yet unmonitored, "reservoir" of carbon on land that is expanding—perhaps forests in the boreal zones?

A carbon scenario unfolds, repeatedly, and in a grand chorus: Plant detritus is consumed by a worm, which releases carbon dioxide, which dissolves into groundwater that flows into river and then ocean, and from there diffuses as a gas back into the atmosphere. Then photosynthesis brings it again into a plant's tissue—perhaps on the other side of the planet. A fallen leaf from this second plant is caressed and consumed by the searching hyphae of a fungus, bypassing the worm this time around. The cycle thus contains both sameness and difference.

Sameness and difference are united through the archetype of the helix. Helixes embody cycles as parts of arrows. The cycles repeat; but at a larger time scale the process, the system, is going somewhere. Each year I celebrate or, more often, ignore my birthday; it comes like clockwork, but I am getting older. Of more cosmic import is the helical path of evolution. Disparate atoms coalesce into molecules, molecules into cells, cells into more complicated cells and organisms. Meanwhile, in each level at each coalescence, new relations radiate, pulling the stream of life outward into all the nooks and crannies of a level and, fatefully, up into an altogether new level.

The "going somewhere"—the arrow part of a helix—need not always attach to evolutionary invention; another grand feat of life is persistence in forward-flowing time. Consider: from the metabolism of a swan to that of Gaia, the main force that dynamic entities must counter is faceless, disruptive entropy, which disperses heat and in general works to dissipate concentrations and complexities that may have been carefully sequestered and built by life. In cells, proteins break down continually, and so, in a vigorous turnover, they must be synthesized anew just to hold the ship of life steady. Many neuronal pathways need a certain frequency of firings to maintain their structure; we thus invigorate memories by remembering them. Over and over, processes at all scales repeat as stones in the foundations of everything. "Before enlightenment, washing dishes; after enlightenment, washing dishes."

Perhaps the most common pattern in the cycles that drive arrows is our old friend the binary. In the pizza parlor pulsed up & down, enter & leave, in & out. As binaries emerge from virtually all the other meta-patterns of space—the sphere and its environment, the two sides of a border, the directions in a tube, the sphere-tube synergy, the center and its peripheral parts—so, too, emerge binary cycles, as circuits of flow connecting the binaries of space.

When the two parts of a binary interact in the intense synergy of mutual causation, they propel the arrow by impelling each other. In electromagnetic radiation (the very adjective contains the binary), electric and magnetic waves create and recreate each other in a self-sustained oscillation that pierces through space in an arrow called light. Republicans pick up on the public unease brought about (they say) by the Democrats; they get elected and rule until the rise of

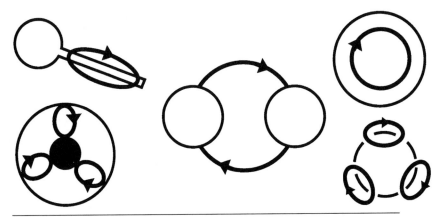

Cycles and the metapatterns of space. *Center:* feedback loop between members of a binary. *Clockwise from upper right:* a circuit within a sphere; to and fro across the border; circulation between center and periphery; back and forth along a tube.

unease in turn is exploited by the Democrats. During a tennis match, each return is empowered, in part, by reversing the momentum of the previous hit, while spectators' heads sweep in unison: no, no, no, no, no. A species of aphid evolves habits on two different plants (sumac and moss) and ends up with a double-host life cycle, with each host giving rise to the stage of aphid that migrates to the other.

In addition to this irreducible structure of binary around which the cycles of nature can work, a psychological side to cycles as binaries should be considered. For example, the shortest traditional Indonesian week, the *duwiwara*, beats with two days, M'ga and P'pat; rituals of the ancient Hawaiians cycled annually between times dominated by either the god Lono or the god Ku. In a persistent theme of western allegorical art, humans rode Fortuna's wheel, rising and falling, enriching and impoverishing; those riding on top of the world were just poised for decline.

As with the binaries of space, binary cycles can aid comprehension; they are tools for thinking and communicating. One way to compre-

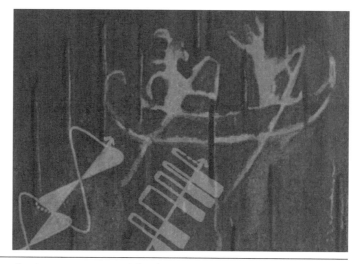

Binary cycles in arrows. *Lower left:* a beam of light has electric and magnetic waves at right angles to each other—one is negative while the other is positive. *Lower center:* U.S. politics is not a contest but a semiregular wave of Republican and Democratic pulses, shown here for this century of presidencies. When canoeing, strokes and returns alternate, plunging and emerging between water and air (Peruvian etching of canoers on birch bark).

hend the myriad spirals that carbon atoms follow through the biosphere is to reduce that chemical movement into just a few binary cycles that capture the essence of a journey through millions of types of organisms, thousands of land biomes, the vast oceans, and layers of rocks in the geologic reservoirs. One such binary cycle entails photosynthesis and respiration, which shunts carbon atoms back and forth between the two great general states of organic and inorganic forms. Another grand cycle is the diffusive shuttle of gas exchange back and forth between air and sea. In still another, carbon falls from the ocean's surface as "marine snow" into the black depths, where much of it is oxidized and returned via the tortuously slow upwelling of water to the surface, as carbon dioxide ready to be photosynthesized again. Considering even longer time scales brings us to the geochemical carbonate-silicate loop, whereby calcium freed from tough silicates on land hitches up with bicarbonate for a river ride to the sea, where it is deposited as white sediments of calcium carbonate. Subducted down the ocean trenches, these carbonates are heat-shattered back into silicates, releasing carbon dioxide gas in volcanic belches. The global carbon cycle encompasses all of these binary cycles, with leaks back and forth across one another.

Cycles not only maintain the larger arrows of persistence (Gaia's metabolism, the tenacity of life on Earth). They also propel the arrows

Binary components of the global carbon cycle. As calcium cycles between silicates (dark) and carbonates (light), carbon journeys between freedom in the atmosphere and the crystalline cage of carbonate. The two rock forms of calcium here lie together on a beach in Hawaii, but usually their formation is separated by thousands of kilometers and hundreds of millions of years.

of change (evolution). Cycles are the pumps that drive the arrows of accumulation, development, progress. Howard Odum, systems ecologist, sees in metabolic systems of all shapes and flavors a generic cycle essential for growth and evolution. These systems dedicate some of the energy and matter they gather to construct more components that do the gathering. This is the classic path of industrial progress. Historian James Beniger points out that a significant portion of the pig iron and then steel manufactured with the power of coal from the mid to late 1800s in America was consumed by the railroads, thus enabling delivery of still more coal and raw materials to the mills—a cycle powering an arrow.

Charles Darwin conceived of a cycle to explain the arrow of evolutionary lineages, of species transforming one to another and to yet another. Variation and natural selection constitute the hep one, hep two, hep one march that underlies the historic pageant of life. Gregory Bateson saw this dance of variation and selection not only in evolution but in the other "great stochastic process"—learning. Like evolution, learning scatters variety and then filters the results into subsequent refinements by judging error and success.

For the Darwinian two-step to work on the lineage of any multi-celled organism, Richard Dawkins contends that the life cycle must pass through a single-celled stage from one generation to the next— which is in fact the pattern (with brief clonal excursions) for all plants and animals whose life cycles are known. This technique for reproduction allows genetic variation, which arises statistically in the chromosomes of a germ cell, to radiate from its origin to affect the whole trillion-cell behemoths that grow to eat, excrete, and copulate.

Cycles are also essential to the arrows of science. In our training years, the work of young scientists resembles the maintenance metabolism of cells and ecosystems. Truth is preserved and even spread by reproducing past results. This flow of duplicate findings between contemporaries and even across generations is one pillar of science: Give me the recipe for an experiment and I will see for myself. Consider, too, the binary system called double description. This coming to the same conclusion by more than one route—verification rather than mere replication—is not strictly a repeated clonon in time, like reproduced experiments. In fact, that the routes to the same place are different is the crux of double description. We in science are always questing for verification and still more validation—or falsification—for ideas and models.

The success of western science—the arrow of scientific progress—indeed owes a great deal to the grand cycle of the scientific method itself. The cyclic nature of the method is evident in the notebooks of Robert Hooke, famed experimenter, microscopist, and rival to Newton at the dawn of formalized procedures for science:

The Method (therefore) of making Experiments by the Royal Society, I conceive, should be this.

First, To propound the Design & Aim of the Curator in his present Enquiry.

Secondly, To make the Experiment, or Experiments, leisurely, and with Care & Exactness.

Thirdly, To be diligent, accurate, & curious, in taking Notice of, & shewing to the Assembly of Spectators, such Circumstances and Effects therein occurring, as are material, or at least, as he conceives such, in order to his Theory.

Fourthly, After finishing the Experiment, to discourse, argue, defend, & further explain, such Circumstances & Effects in the

Power cycles in the arrows of civilization. Wheels turned by pressurized steam, wind, and diesel fuel generate electricity, grind grain, and ply the Mississippi River.

preceding Experiments, as may seem dubious or difficult: And to propound what new Difficulties & Queries do occur, that require other Trials & Experiments to be made.

Scientists thus live on spirals of ignorance and insight. We found oxygen, now let's go and discover plutonium. We know this enzyme, now let's find the gene that makes it. Discovery is a bicycle of question and answer.

After carrying through experiments and raising a new round of questions, scientists are still not done. They need to publish. Hooke himself included a fifth step in his "Method of making Experiments": the would-be natural philosopher must "register the whole Process of the Proposal, Design, Experiment, Success, or Failure." This continues today in the highly formalized arrow of research publication, which itself contains subcycles.

First come the interminable rounds of one's own drafts, leading finally to submission to a journal of what we call a "paper." The technical editor then sends it out to two or more reviewers competent in the field. When the reviews return and are considered, the editor may accept the paper outright (unusual), flatly reject it, or encourage the author to resubmit—once the minor annoyances or even major gaps found by the reviewers are addressed. How it stings to get a bad review! This can send one back to the drawing board, lab bench, or computer. To resubmit, one must justify the changes and hope that the editor, the ultimate "center," will find them worthy. But it is not uncommon that even after a wrenching revision the paper requires still another round.

On a larger scale the arrow of publication turns into a cycle that then propels the grand arrow of scientific progress. One scientist gets an idea by reading a paper already published by somebody else; he or she runs with the idea and produces a new paper—one that cites or acknowledges the first. The scientific arrow of discovery and progress viewed at an even larger scale reveals the rises and falls of paradigms, which may actually be cycles within the larger arrow of the evolution of knowledge.

Arrows within arrows, cycles within arrows, arrows within cycles, and cycles within cycles: these holarchies in time are present in the patterns that make and surround us. They evince harmonies and changes, repetitions and variations, at scales from atomic vibrations to galactic carousels, from the shimmers of nucleotides to the sighs of kings and the groans of societies. It is all a totality that, to me, most resembles music.

Music of the Cycles

Is the astrophysical universe an arrow or a cycle? Will the community of galaxies expand forever and lapse into darkness when all useful energy is spent? Or will a retraction at one point begin, an inhalation that could fuel another big bang of a universe?

Is biological evolution an arrow? or cycles of random walks, pure happenstance? Is the sequence from fish to amphibian to reptile to mammal and bird just that: a lovely, but meaningless, stream of change? Or is evolution going somewhere?

What about civilization? Is it riding an arrow of something called progress, a concept that often seems to be partner with time itself? Or are the rises and falls of the walls of Jericho, Babylon, and Florence like the bloomings and fadings of so many flowers?

One early morning, before any other tourists were about, I visited the ruins of what had been the only round temple at Delphi. It was situated near the site of the gymnasium, where Hellenic boys and girls trained naked for the games. Conifers now scattered the site with shade. The columns of this round temple were mostly toppled, and the presumed cupola cap long gone.

The only glory remained in the half-buried fragments from the mysterious round edifice. They spoke of intricate craft from chisel to sculpture, in whirls and words so finely rendered that one is sure that the makers believed their work immortal. We today build World Trade Tubes and Capitol Spheres not at all with their disassembly in mind. Those who build in marble or steel and concrete build for the arrow not the cycle.

A Hindu tale imparts a stern lesson for those who worship the arrow and forget the cycle. Indra, most glorious ruler and builder of the most magnificent palace, is visited by Krishna, disguised as an innocent lad. Krishna arrives with a parade of ants to drive home his point that even Indra is just one clonon of an infinity of Indra-cycles. The lad points to the ants and cries, "Behold, former Indras all!" Shelley expressed the same sentiment in his poem about the once-towering statue of Ozymandius, reduced to rubble. Even though we may think we are riding a one-way road into an ever-enlivening, cola-commercial future, the legions of fossils in museums and the now-homogenizing landscape ought to give us pause. For the organisms and taxa extinguished, it matters not whether the bludgeon is a comet from space or one klutz of a species here on Earth.

Are these thoughts too dark? Well, to follow the balance recommended by wilderness advocates, there are times for indignant activism and other times for just hiking and enjoying the wild nature that remains. Indeed, all the metapatterns of time can be found in the song of one bird. Starlings are sophisticated songsters. They learn from other birds (and even mammals) wherever they live, and the resulting regional differences in the starling repertoire serves as a study in the transmission of nonhuman culture.

The main features of starling songs are the cycles of lush melody, in syncopated bursts of notes and trills. These patterns at times develop over many cycles, and thus move along arrows. Sometimes the shift is a dramatic break into a new pattern. Sometimes patterns, at a higher level, are repeated. From the most elemental level—the binary beat of sound and silence, of note and non-note, of sphere and environment in time—the holarchy of starling song is formed by perhaps four or five levels of patterns encompassing patterns.

Many organisms of course have fewer levels in their songs than have the starlings. But there are surprises. I remember putting my ear to a rock crevice with hidden cricket. At first I thought that all the cricket could offer was a simple, repeating pattern of two chirps. Suddenly it started jamming with its code in a way as entrancing as a jazz drummer.

After birds and crickets, give ear to the complex holarchy of time patterns in a Beethoven symphony. Disregarding the simultaneous parts of the orchestral sections, I count about three more levels than that of the starling. Starting with the symphony itself as the most encompassing level and moving inward, a handful of movements sound out a second level. Further analysis discloses more parts within parts: what are called the portion, the material, the bars, and the simple substructure within bars (for example, 8 as 2×4; 11 as $2 \times 5 + 1$). Finally, to these I add the figure & ground binary of note and silence, bringing the levels to at least seven.

Somewhere between starling and Beethoven (or at least between starling and cricket) booms the time holarchy called rock-and-roll.

Cycles, breaks, and arrows in music. Nest-site song of a European starling in New Zealand, with frequencies 0–8000 Hz. Sonata for Violoncello and Piano in A Major by Beethoven; this portion of the manuscript shows composing in process.

Rarely are rock songs reduced to a single chord. But it is surprising how satisfying two chords can be. A simple binary cycle starts the rocking, back and forth, the swaying of the sphere of trance in its alternating arrows of change and return, of distance and home.

A rock composer will usually opt for a third chord. Examples include the riff of "Louie Louie" and the classic twelve-bar blues progression. Beyond the binary, in proper hands the three chords have remarkable power. Even more chords can be added, of course, but the result is almost always a small number arranged to repeat, over and over and over. In "Louie Louie" the chord sequence repeats throughout the entire song. Yet even this simple tune alternates verses and choruses, creating an even more encompassing level. Many songs—say, by the Beatles—still more fully differentiate verse and chorus with unique chord patterns for each. Canonically, each verse uses a common melody, but, on another level again, different words. So as a whole the sequence of verses weaves a tale or travels an emotion. The choruses, on the other hand, are utter clonons—nearly and almost always repeated exactly, except for an interjected hoot or two. Thus verse and chorus mix as a grand binary, out of which comes a stew of cycles and arrows.

The standard rock anthem, after a couple rounds of verses and choruses, breaks into yet another level of pattern. In jazz this entire next period is literally called what it is—"the break." It is the time for soloing, the time for improvisation, the time for magic. Here, in its best, the musicians step off the printed page and into a creative dream.

"Let's jam."

Where we might go, no one knows. And the melodic improvised arrow (atop the cyclic background beat and riff) might wend on forever except for some preplanned sequence that will be played at the discretion of a designated center, thus signaling to all that it is time to head home. The arrow then closes into a great cycle, and the song reenters the original territory of the verse and chorus subcycle for one or two last rounds.

The verse and chorus structure can also be seen in a rhetorical device of great speeches, poems, and rituals. In the Bible each day of creation ends with the intonation of a cosmic conclusion: And God saw that it was good. When Martin Luther King, Jr. asks for justice, we hear: How long? not long . . . How long? not long.

A koan in Zen will be asked over and over. Hear the Hopi chant at the beginning of the film of the same name: Koyanisqaatsi . . . Koyanisqaatsi . . . Koyanisqaatsi . . . Koyanisqaatsi . . .

Surely somebody somewhere right now is listening to the Beatles: All you need is love . . . all you need is love. Somewhere Jean Houston is evoking yet another group trance: Imagine yourself in an activity you'd like to improve . . . now do it better than ever before . . . now do it as a master . . . now be it! Somewhere in Africa at a naming ceremony, women are singing: deeply, deeply the dye goes into the cloth . . . deeply, deeply the dye goes into the cloth. Somewhere a shaman sings to the north wind: Fox sits with me . . . skunk sits with me . . . datura sits with me. Somewhere the sacred ball is tossed to the east, then the south, then the west, and with each toss the swift run to catch the sphere of life.

Life is an arrow filled with and propelled by cycles, habits. Life is a star that rises and falls across the horizon. Life is decisive breaks. Life is drones from a sitar, chimes from a Balinese gamelan, cries from a Sufi minstrel, trills from a mammoth bone flute, an hallucinogenic mosaic from a shaman's drumming, echoes from a stone fly drumming on a rock along a river.

Twenty years ago I first came across mention of Jean-Jacques Rousseau's revelation at Lake Bienne in Switzerland, by way of historian Kenneth Clark. Ten years later, while studying for Ph.D. entry exams (and with not a minute to spare), I was nevertheless drawn into Rousseau's *Confessions*. Although he spoke wondrously of his few months in the sweetness of doing nothing on a tiny island in the middle of the lake, I did not see the revelation. The idea haunted me that I might have missed something, especially after revisiting Clark's description. What did Rousseau discover?

Several years later, returning to the *Confessions*, I again saw nothing profound. Then only recently, when I stumbled on Rousseau's book called *The Reveries of the Solitary Walker*, I finally found it. The fifth "walk" is the gem. On Lake Bienne, Rousseau came to cherish a glorious habit.

> When evening approached, I would come down from the heights of the island and gladly go sit in some hidden nook along the beach at the edge of the lake. There, the noise of the waves and the tossing of the water, captivating my senses and chasing all

other disturbance from my soul, plunged it into a delightful reverie in which night would often surprise me without my having noticed it. The ebb and flow of this water and its noise, continual but magnified at intervals, striking my ears and eyes without respite, took the place of the internal movements which reverie extinguished within me and was enough to make me feel my existence with pleasure and without taking trouble to think.

Rousseau extolled this "sentiment of existence . . . of contentment and of peace which alone would . . . suffice." To enter such states, he found that one "must be favorable to them, as must be the conjunction of the surrounding objects. What is needed is neither absolute rest nor too much agitation, but a uniform and moderated movement having neither jolts nor lapses."

Hiking in the autumn Precambrian highlands of lower New York, reaching one of our "spots," now aflame with red blueberry bushes, Connie says she'll ignore the book she brought and just be off to find a tree for her communion. I know she will smell the warmth of sweet decay, and mostly just lie and watch the leaves above—yellow and orange on blue—jostle in the wind. Ah, just to sink into anything in nature with that right meter of motion, that transporting mix of repetition and variation, of cycles in arrows of larger cycles.

Okay, I say, see you in a couple hours.

Late in life, despite a litany of troubles, nothing could prevent Rousseau from revisiting Lake Bienne—though he did not do so physically. He wrote of "transporting myself there each day on the wings of my imagination and enjoying for a few hours the same pleasure as if I were still living there." As Rousseau cycles his memories, so can I in turn.

At the start of the trail to the cliff dwellings at road's end, a woman reads to her child from the pamphlet that guides all visitors through numbered stops on the circuit. First they will cross the wooden bridge over the river, then huff and puff up the dirt path to the dwellings, and finally switch back around a return loop down to the bridge, thus completing a circuit taken by a hundred thousand others that year.

"The cliff dwellings are eight hundred years old . . . no one knows why they were built or why they were abandoned so suddenly." Had the wild game dwindled, a drought lingered, the firewood become depleted? Or were myths the cause?—we suspect that the myths of the ancient southwesterners, like those of their descendants, revolved around wanderings, settlings, and then renewed wanderings. Did these tales prompt their moves, or at least provide an optimistic face to an environmentally caused trauma?

Many visitors honorably save fifty cents by returning the pamphlet for the next curious hiker of the trail. Written by someone they will never know, the words of one go into the minds of others, again and again. Each time they engender a unique blend of questions, jokes, musings, silences. The dwellings themselves are also books, patterns printed in stone, stacked as borders against turbulence and tempera-ture, within a fractal tube-work of valleys, under the great blue dome. We see the walls: ancient Americans built them, and both living and dead have touched them. The building, the living, the leaving, the marveling—do these acts speak of a cycle? Or arrow? Or helix?

Bicycling home, the folk melody from the choral movement of Beethoven's Ninth smoothes the route along the river. This is one of the now hundreds of such sojourns—an evening ritual—clearly cyclic, and yet each unique and each taking me into the land, and so an arrow too. Singing simultaneously with pedaling; singing simulta-neously with the life cycles of the plants that can count on the river and their upland kin that cannot.

But it is the rainy season: many now bloom along the road. Green sheets unfold from slender tubes rising skyward, at whose ends radi-ate rosettes of more sheets with yellows, whites, and purples. I stop to catch the rich drafts of aroma from fresh blossoms of hand-size Sacred Datura as they open to the twilight, designed to tempt the sphinx moth, the hummingbird of the night. At the flower's center resides the great binary of pistil and stamens, producing the tiny spheres of eggs and pollen that merge to make seeds fill and grow. Spheres come from tubes, binaries dissolve into trinaries, cycles emerge from arrows—all the metapatterns shout through one single, perfect, and sacred plant.

We each began, attached to the wall of the womb, sunk into and wrapped by the placental border, later connected by a tube, within

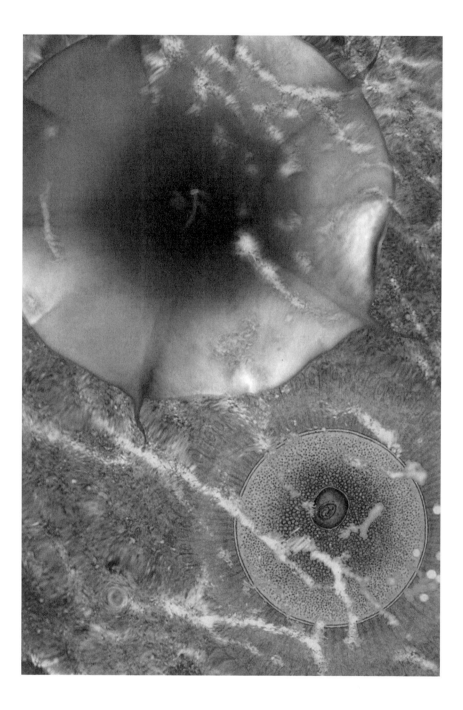

Flower of Sacred Datura and human ovum.

mother, within society, within a planetary biosphere, the center and the periphery, growing on an arrow with cycling heart readying for the break of birth. And all this, from a majestic lineage of beginnings.

As we can always remember, can always revisit, can always recall— "As spheres floating simple, as life, began We."

Epilogue

What Are Metapatterns?—Revisited

What then are the metapatterns?
Perhaps look to the wind for an answer; listen to the wind.
It whispers and bellows, but I hear no words.
Then try the moon; go ask the moon.
It waxes eloquent, but again not in words.
Then seek out the grape; go ask the grape.
It rounds out the conversation; ah! . . . not with words.

Acknowledgments

Many have given specific and essential aid to this project since 1987. First I wish to thank those friends and colleagues not expressly mentioned here who provided advice and factual details over the years.

I am indebted to those who, early on, conferred written comments on Columbia University Press or New York University, thus helping to move this project along: O. Roger Anderson, Ann Blackwell, Hatice Cullingford, Alexander George, George Sugihara, and Jo Yohay.

Especially important in the final stages were Stuart Gaffin, John Richards, and Luke Wallin. As professional scholars and interdisciplinary thinkers, they read the entire manuscript and provided detailed comments to the press. Their reactions and suggested changes were crucial. I was always able to count on Martin Hoffert, my long-time mentor on climate and the carbon cycle, for encouragement and enthusiasm. Also, Orren Champer read and wrote marginalia on the entire draft; I thank him and apologize for not heeding his advice to "wait until you are sixty to write this book."

In 1985 Edwin Dobb, then senior editor at *The Sciences*, reworked my piece on spheres, which later proved to be a key part of the book proposal. I thank him for his friendship and writing advice over the intervening years.

A big step was taken over lunch in Manhattan's East Village in the spring of 1987. Ed Lugenbeel, executive editor at Columbia University Press, talked me into submitting a book proposal. He made it sound so easy—little did I know how much time during the next seven years this project would deliciously consume. To him goes my immense gratitude for his patience, for his immediate responses to chapters as they trickled in, for his ever-growing support that would rally me on to more, and for what has turned into a rewarding friendship. Everyone at the press has been great to work with—energetic, professional, and full of ideas.

Tim Binkley, for many years chair of humanities and sciences at the School of Visual Arts, supported the teaching of my oddball patterns curriculum as a science elective. I encouraged students to show that they grasped the science I was presenting in the way they knew best. Imaginative posters and other visual projects were the results, which often verged on performance art. I have happy memories of those wild times and thank the hundreds of students.

Lyn Hughes, photographer extraordinaire, visual thinker, and spirited friend, generously allowed me to use many of her photographs, both entire and as parts of collages, and thereby to incorporate special moments from our travels.

I thank New York University for granting me two essential leaves. First, a Junior Faculty Fellowship in the spring of 1989 permitted me to launch the actual writing. Then a sabbatical in 1991–1992 provided time for continued research and half of the initial draft. Even when in New York, I could not imagine a better place to be than NYU for pursuing interdisciplinary scholarship. This included the opportunity to teach a semester in New York University's graduate program in liberal studies, and I thank those students also.

Musical thanks go to Liz Story, Lyle Mays, and Anton Bruckner.

Finally, for Connie Barlow words are not enough. She is the partner in this book, as in my life and love. Her hands, heart, eyes, and mind were integral to every level, from reading and annotating the first words I produced on the North Carolina shore, to officially editing the manuscript for the press (not to mention instigating some of the humor in the figure captions). An exceptional writer who has taught me much about words, philosophy, nature, and the good life, Connie has been vital to my dreams and inspirations under full moon nights.

Notes

The number of notes and references was limited by mind, time, and (mostly) space. The many lists in the text, for example, could have supported a note for each item. I have cited all quotations (except some from popular culture) and whatever information I judged to be arcane or not easily retrievable. As a scientist accustomed to citing thirty papers in a four-page submission, I found the job of pruning difficult. I apologize for the necessary omissions and for the idiosyncratic nature of some entries. I chose an informal (superscript-less) style of notation in the interest of aesthetics.

1. SPHERES

p. 4, **Melasmatosphaera:** Photo 14-2-V and ch. 14 in Schopf 1983.

p. 4, **Hawking event horizon:** At a certain radius from a black hole, pairs of particles are spontaneously created; one is drawn inward and one radiates away (Hawking 1988).

p. 4, **only the very small bodies . . . :** I ignore the galaxies, which come in various shapes: spindle-shaped, flat and spiral-armed (ours), and some indeed spheroidal.

p. 6, **scanning tunneling microscope . . . :** See the atoms in the second figure of this chapter.

p. 6, **grape-moon koan:** Many koans are entirely verbal ("What is the sound of one hand clapping?"). The inked circle of Zen art—the enso—was used as a visual koan by Hakuin (Addiss 1989, p. 124).

p. 9, **water jug:** This story comes from the formation of the short-lived Wei-yang sect of Zen Buddhism, more than a thousand years ago (Dumoulin

1963, p. 107). A modern Zen message about names is the joke, "Call me anything you wish, but don't call me 'late for dinner'."

p. 10, **Story Musgrave:** Our talk was at NASA's Johnson Space Center, during summer 1987. Multi-talented Musgrave headed the team that repaired the Hubble Space Telescope in late 1993.

p. 12, **minimized surface area . . . biological advantages:** D'Arcy Thompson (1971/1917) masterfully elaborated this idea.

p. 13, **buckyballs are incredibly strong . . . :** *Science* (1991) 252:646.

p. 14, **Harry Kroto . . . :** The story, including codiscoverer Richard Smalley's own "eureka experience," was reported by Taubes (1991).

p. 14, **. . . largest clear-span, self-supporting enclosure . . . :** In Fuller 1981, p. 389.

p. 14, **Roman Pantheon . . . :** Whatever went on in those unknown activities, architectural analysis has determined that it involved great crowds of people—unlike, for example, the interior activities of the rectangular Greek temple, the Parthenon (Giedion 1971, p. 148). For an inspiring discussion of the Pantheon and other domes in a cultural context, see Hardison 1989.

p. 14, **Roman arches and domes were basically compressional . . . :** I have ignored the not-minor issue of how they contained the potentially bursting tension at their bases.

p. 17, **edible part of the outer rind:** This so-called caruncle contains "oily material that is much prized by ants, which gather the seeds and transport them to their nests." Such underground seeds sans caruncle have better germination statistics than seeds left above ground (Moore 1992).

p. 18, **"magic molecule" and "Molecule of the Year":** Names bestowed by *Popular Science* (August 1991) and *Science* (1991) 254:1705.

p. 18, **"The game as it is played today . . . ":** Black Elk's quotation in Brown 1989, p. 127. For smoothness, I have deleted minor phrases and did not use ellipses.

p. 18, **cosmic egg:** The concept existed in India, Egypt, Persia, and Greece (Smith 1971, p. 77).

p. 18, **Neolithic . . . birdlike goddesses:** In B. Johnson 1988.

p. 20, **mortuary mounds . . . dome tombs . . . :** See Smith 1971 and Giedion 1971.

p. 20, **"domes of heaven":** The origins of this iconography in the ancient world and its development through Christianity are treated at length in Lehmann 1945. See also Smith 1971 and Lethaby 1975, ch. 10.

p. 22, **The Sphear-like forme of his Crowne . . . :** Quoted in Soellner 1984.

p. 22, **Temples of Equality:** See Rosenau 1976.

p. 22, **"Creator compounded the world . . . "**: Plato, in Hutchins 1952, vol. 7.

p. 23, **". . . the universe is spherical . . . "**: Copernicus 1543.

p. 23, **sphere is the "image of God . . . "**: Kepler, in Hutchins 1952, vol. 16, p. 853.

p. 25, **circle of friends:** Relatedly, Diane Ackerman (1990) cites a number of metaphors, concepts, and social customs that use "rings."

p. 26, **Maurice Strong . . . :** From *Natural History Magazine,* June 1992, p. 6.

p. 26, **"Behold today Rattling Hail Woman . . . "**: Black Elk's quotation is in Brown 1989, p. 137. I changed "Thy earth" and "Your world" to "the earth" and "the world."

2. SHEETS AND TUBES

p. 31, **research voyage:** I was aboard the *RV Knorr* for the program Transient Tracers of the Ocean, Equatorial Atlantic Expedition, 1983.

p. 31, **. . . an area about four times that of the continents:** Mankiewicz 1991.

p. 31, **twenty more times the amount of skin:** Greulach 1973, p. 3.

p. 33, **"differences that make a difference":** Definition of information, Bateson 1979.

p. 36, **Wainwright . . . :** See Wainright 1988. For quotations, pp. 17, 96.

p. 39, **large blood vessels . . . small blood vessels . . . :** LaBarbera (1990) employs the terms "long-distance transport" v. "exchange sites." Vogel (1988) provides an overview of fluid dynamics in biological tubes.

p. 40, **aboriginal artists of Australia . . . :** Sutton (1988) showed the beautiful "dreamings" and discussed the circular roundels and connecting paths.

p. 40, **Tree of Life . . . :** In Halevi 1988, pp. 5, 40, 65.

p. 41, **Katmandu . . . stupa:** A striking image is in Campbell 1974, p. 22.

p. 42, **shaman's path of transcendence:** A first and requisite experience to learn shamanism is taking the Tunnel into the Lowerworld (Harner 1990). See also Halifax 1982.

p. 43, **During Sioux rituals . . . sacred pipe . . . :** See Brown 1989, especially pp. 39, 76, 78.

p. 43, **tree of Buddha . . . cross of Jesus:** Symbolic connections suggested by Joseph Campbell in various works.

p. 43, **Ohio River culture labored . . . :** If outstretched, the mounded serpent (with egg in mouth) would be 400 meters long (Morgan 1980).

p. 43, **Very young children . . . the tadpole figure:** See Freeman 1987.

p. 43, **"far more deeply interfused"**: Phrase from Wordsworth's poem "Lines Composed a Few Miles Above Tintern Abbey," 1798.

p. 43, **marble Galileo . . .** : The sculpture in Florence is a commemorative structure with his tomb.

p. 45, **diagram . . . Earth's carbon cycle . . .** : The picture with circles and lines can be drawn with the tools (classically limited to two) used for Euclid's geometry proofs and constructions: compass and straightedge.

p. 46, **generic modeling programs . . .** : I refer to the object-oriented programs or languages, with graphical interfaces. Many are available and are being improved before our eyes.

3. BORDERS

p. 51, **genetic disaster . . .** : See Blerkom and Henry 1991.

p. 53, **rings of . . . Babylon** and **town . . . Zaun:** Schneider 1963, pp. 32–33.

p. 53, **Chinese word for city . . .** : Hook 1982, p. 444.

p. 54, **Legend has it that Meng Jiang . . .** : Information on the Great Wall from Fryer 1975.

p. 54, **Cambodian . . . Ramayana:** Performed by the Classical Dance Company of Cambodia, October 1990, at the Joyce Theater in New York.

p. 54, **Jade Gate:** Through it Lao Tsu left China near the end of his life. But just before, he was talked into writing his wisdom, now known as the *Dao de Jing* (Fryer 1975, p. 28).

p. 55, NASA **project . . .** CELSS **. . .** : For overviews see Ming and Henninger 1989.

p. 55, **nearly closed to matter fluxes:** Implications for Earth and the Gaia hypothesis are discussed in Barlow and Volk 1990.

p. 57, **walls and bridges:** This binary in the metapattern of borders may have been recognized by John Lennon in the title of his album of the same words. Ken Wilbur (1981, p. 25) also saw the binary: "so-called 'dividing lines' . . . join and unite just as much as they divide and distinguish."

p. 57, **spectrum of visible edges from crisp to fuzzy:** Zerubavel (1993) establishes a psychological spectrum of how humans make distinctions, with concepts for mind as "rigid," "fuzzy," and "flexible."

p. 57, **"functional" borders . . .** : An excellent distinction between functional and structural boundaries is in Allen and Starr 1982, pp. 67–93.

p. 58, **an ecosystem lacks . . . a membrane:** Allen and Hoekstra (1992) discuss difficulties in conceptualizing ecosystems because of an absence of discrete boundaries.

p. 60, **we decorated our skins . . .** : Flugel 1950, pp. 16, 71; Payne 1965, p. 1.

p. 60, **"the doors of perception":** Huxley (1954) used this phrase from William Blake's poem "The Marriage of Heaven and Hell," 1790.

p. 61, **"container metaphor":** In their eye-opening book, *Metaphors We Live By*, Lakoff and Johnson (1980) named this metaphor and provided examples for the sentences about Janice.

p. 61, **Moksha:** Huxley 1977.

p. 62, **container metaphor in . . . prefixes . . . :** Inspiration for my examples of prefixes and, in the next paragraph, for border metaphors in daily expressions came from Lakoff and Johnson 1980. A host of other examples and insights along these lines is in Zerubavel 1993.

p. 62, **convoluted rocks:** The "eyes" of these *T'ai-hu* rocks are also called "leaks" (Keswick 1986, pp. 75, 161).

p. 63, **Sioux's rites:** See Brown 1989.

p. 65, **Isis conceives . . . :** For the story see Campbell 1974.

p. 65, **Iron John . . . :** For the story, full of themes with barriers and crossings, see Bly 1990.

p. 65, **hero with a thousand faces:** Phrase from the title of Campbell 1973.

p. 65, **Copernicus . . . :** Quoted in Hutchins 1952, vol. 16.

p. 65, **Wheeler . . . :** So described by *Scientific American*, June 1991, p. 36.

p. 65, **Peattie . . . :** Quoted from his book *Flowering Earth* in *New York Times Book Review*, Dec. 31, 1991.

p. 66, **Liu Ling . . . :** See Keswick 1978, pp. 73–74.

p. 68, **tien yuan, ti fang . . . :** From Keswick 1978, p. 199.

p. 69, **"Robin redbreast . . . ":** From William Blake's poem "The Auguries of Innocence," 1804.

p. 69, **Confucius declared that . . . :** In Keswick 1978, p. 198.

p. 70, **"You ask me why . . . ":** Li Po poem from Keswick 1978, p. 6.

p. 70, **"The wild deer . . . ":** From William Blake's poem "The Auguries of Innocence," 1804.

p. 71, **"The atmosphere is not merely . . . ":** quotation from Lovelock 1979, p. 10.

p. 71, **. . . created by . . . the biosphere . . . :** To what extent the atmosphere, for example, was "designed to maintain a chosen environment" is still a matter of scientific controversy. I have personally worked on quantifying the effects that the evolution of life has had on atmospheric carbon dioxide (Volk 1987a, 1987b, 1989; Schwartzman and Volk 1989). For an overview of the Gaia debate, see Barlow 1991 and Barlow and Volk 1992.

4. BINARIES

p. 74, **sulfur & mercury . . . generation of metals:** Eliade 1971, p. 48.

p. 75, **James Watson decided . . .** : Watson 1968, pp. 171, 175.

p. 76, **Carl Jung showed . . .** : Jung (1978) focused on alchemy, but did include the yin & yang in his amazing analysis.

p. 78, **deaf babies "babble" with their hands . . .** : *New York Times*, March 22, 1991.

p. 78, **Descartes claimed . . .** : I have read that his binary may be apocryphal (and variously expressed). The socially remembered pattern (even if altered) is worth examining.

p. 79, **"The key is balance . . . "**: Quotation from Gore 1992, p. 367.

p. 80, **Jean Houston claims . . .** : In Houston 1992, p. 8.

p. 80, **competition . . . within a more encompassing level of cooperation:** For social implications, see Eisler 1987 and Eisler and Combs 1992.

p. 80, **Richard Dawkins . . . "arms races":** Dawkins 1987.

p. 80, **continual vying is itself the desired end:** Deborah Tannen (1990) enlivens this issue in her chapter section entitled "A good knockdown drag-out fight."

p. 82, **Ruth Benedict:** Her 1940s notion of cultural synergy is known from lecture notes posthumously shaped into essays (Benedict 1992).

p. 82, **Fuller:** His explication of synergy can be found in virtually any of his works, for example, Fuller 1981, 1992.

p. 82, **reductionism & holism:** Both are necessary for science (Bonner 1988, p. ix).

p. 85, **binaries haunting the very roots of thinking:** Stephen Jay Gould makes some poignant comments on dichotomies in thinking (Gould 1987, pp. 8–10).

p. 85, **double description:** The example of adding a column both up and down is mine, but Bateson's own examples are a must-read, ranging across Shakespeare, neurons, and explanation itself (Bateson 1979, ch. 3). Also worth a close look is Bateson and Bateson 1988.

p. 86, **Arcosanti:** I worked there in 1972, and in 1990 attended festivities for the twentieth reunion. Go visit—just off I-17, exit 262, about sixty miles north of Phoenix.

p. 89, **Carol Gilligan has discovered . . .** : Gilligan 1982.

p. 89, **Deborah Tannen has observed . . .** : Tannen 1990.

p. 90, **two hands have been a rich source for parallels:** Konner (1990) tracked the history of this subject.

p. 90, **stringing good to up and bad to down:** Material in this paragraph was drawn from Lakoff and Johnson 1980 and Johnson 1987.

p. 91, **Mayan leaders piercing their tongues . . .** : Schele and Freidel 1990.

p. 91, **Sioux shamans ripping their chests** . . . : Brown 1989, ch. 5. (And not just shamans!)

p. 91, **"Perhaps we are divided . . . "**: Quotation from Margulis 1992. The five kingdoms are elucidated in Margulis 1988.

p. 92, **Another way beyond the binary:** Huxley (1968, p. 35) noted the "blessed experience of Not-Two" and the "reconciliation of yes and no," in his utopian novel, *Island.*

p. 93, **"One plus two equals zero . . . "**: From a lengthy, revealing dialogue in Bancroft 1991, p. 66.

p. 94, **trios of goddesses galore** . . . : These and many more are found in McLean 1989. See also Houston 1992.

p. 95, **Yudhishthira, now old and journey-wearied** . . . : A moving cinematic treatment of this seemingly paradoxical ending is in Peter Brooks's *Mahabharata*—rent the video.

p. 95, **"bisociation":** This theory for the experiences of "haha" and "aha" was formulated by Koestler (1979).

5. CENTERS

p. 100, **"I gave orders . . . "**: Louis XIV's quotation in Rule 1974, p. 19.

p. 101, **polyvertexia of flexible** . . . : Fuller 1992, pp. 217–218.

p. 102, **"Each atom of matter . . . "**: From *O.E.D.* entry for "nucleus."

p. 105, **O. T. Avery and colleagues** . . . : A splendid history of the "transforming principle" is McCarty 1985.

p. 109, **"Though a king . . . "**: Purchas, quoted in Prete 1991.

p. 110, **not meant to slight a dispersed . . . component** . . . : The distributed aspect of the bee hive is precisely what Kevin Kelly (1994) emphasizes in his chapter "Hive Mind," a theme for his "rise of neo-biological civilization."

p. 112, **"not procured by any Tyranny . . . "**: Joseph Warder's *True Amazons or Monarchy of Bees*, quoted in Prete 1991.

p. 117, **"God is the infinite sphere . . . "**: Wind 1967, p. 227.

p. 118, **Chuck Berry . . . "president of rock-and-roll":** Concert poster in New York City, early 1990s.

p. 118, **O.E.D. . . . "king of dictionaries":** Barnes and Noble ad, 1990s.

p. 118, **"misconception that cosmic supremacy . . . "**: Fuller 1992, p. 264.

p. 119, **advocates of dispersant philosophies:** A creative and expressive proponent of dispersion is Kelly (1994). A theory of consciousness with the dispersed control of "multiple drafts" is elaborated by Dennett (1991).

p. 119, **Gaia:** The key works are Lovelock 1979, 1988.

p. 120, **shifting all to suit our needs . . .** : Allen and Hoekstra (1992, ch. 9) described an entity they call the anthropogenic biosphere, which now must manage ecosystems in states away from internal equilibrium. Stock (1993, p. 35) contends "Metaman *will* remake the biosphere."

p. 120, **Foreman . . . Turner . . .** : Their parley appeared in *Harper's Magazine*, April 1990, p. 48. For details on the similarities between human culture and tissue malignancy, see Hern 1993.

p. 120, **"As he is master everywhere . . ."**: Mme de Maintenon quotation from Rule 1974, p. 86.

p. 121, **"If one banishes . . ."**: Diderot, from the *Encyclopédie*, quoted in Williamson 1988.

p. 121, **"The essence . . ."**: Spinoza, quoted in Durant 1943, p. 129.

p. 123, **"My house is the red earth . . ."**: Harjo 1989, p. 2.

6. LAYERS

p. 126, **Bruno . . . and other memory wizards . . .** : See Yates 1966.

p. 127, **"Nature works by steps . . ."**: Quotation from Bronowski 1974, pp. 348–349. For his full explication of stratified stability, see Bronowski 1970.

p. 127, **The rise of complexity through layering:** Herbert Simon (1981/1969) invented an oft-quoted parable of two watchmakers to demonstrate the links between stability, complexity, and layering.

p. 128, **emergence of wholes from parts . . .** : See Casti 1994; Cohen and Stewart 1994.

p. 128, **Mice are common, owls rare:** See Wilson 1992, pp. 35–37.

p. 129, **Arthur Koestler called attention . . .** : Koestler 1967.

p. 130, **nested . . . and non-nested hierarchies:** Allen and Starr 1982; Allen and Hoekstra 1984.

p. 130, **coined by Koestler . . .** : Koestler (1979, ch. 1, "The Holarchy") was ambiguous about an exclusive separation between the terms hierarchy and holarchy, in contrast to my usage. But when referring to "the grand holarchies of existence" he clearly meant the nesting of parts into larger wholes.

p. 131, **holarchies and hierarchies have other significant contrasts:** The best description of the fundamental differences between nested and non-nested systems is in Allen and Hoekstra 1984. If you can't find the article, contact Tim Allen at the University of Wisconsin. In the meantime, see Allen and Hoekstra 1992, pp. 31–34.

p. 132, **"The division of the perceived universe . . ."**: Bateson 1979, subhead 5 of ch. 2.

p. 132, **Polonius . . . Hamlet . . .** : *Hamlet* (Act 2, Scene 2).

p. 132, **organisms are at once contained . . . :** A main theme of Allen and Hoekstra 1992.

p. 133, **What is a cell to the chicken? . . . :** Jumps between holarchies fall under the heading of logical types, as emphasized by Bateson (1979, and other works).

p. 134, **Mary Catherine Bateson raises a call . . . :** Quoted from an interview by William Irwin Thompson in *Annals of Earth* (1990) 10, no. 3.

p. 134, **The Powers of Ten . . . :** Vol. 1 of *The Films of Charles and Ray Eames*, Pyramid Films and Video, Santa Monica, Calif., 1989 (original version made in 1968).

p. 137, **. . . complexity theorists:** See the absorbing overviews by Lewin (1992) and Waldrop (1992).

p. 137, **Volvox and sea lettuce:** Their cell types and those of other mentioned organisms are from Bonner 1988.

p. 143, **Living Systems:** Title of Miller 1978.

p. 144, **Robert Pirsig described . . .:** Pirsig 1985, pp. 63–64.

p. 144, **three . . . detrital nitrogen webs . . . :** Moore and Hunt 1988.

p. 144, **White Cloud . . . :** *New York Times*, May 6, 1993, p. D1.

p. 145, **Management theory recommends . . . :** Spencer Weart, pers. comm.

p. 145, **"The magical number seven . . . ":** Miller 1956.

p. 145, **William Calvin predicts . . . :** Calvin 1990.

p. 147, **properties . . . for generic alphabets . . . :** Logan 1986, p. 151. He also pointed out the alphabet nature of numbers.

p. 148, **Hubert Reeves . . . :** Quotations from Reeves 1991, pp. 49, 220.

p. 149, **"As scientists pursued . . . ":** Quotation from Logan 1986, p. 208. If I could, I would reprint Logan's chapter 13, called "Print, the Alphabet, and Science," which I mined for ideas, particularly in the next paragraph, including the quotation "repeatable uniform elements . . . " (p. 207).

p. 150, **Karen Horney saw . . . :** Horney 1945.

p. 151, **"Philosophy is written . . . ":** Galileo, quoted in Logan 1986, p. 200.

7. CALENDARS

p. 153, **Imagine . . . no memory:** Voltaire imagined the amusing consequences of this actually happening to everyone in his 1775 essay, "Memory's Adventure."

p. 155, **spatialization of time:** A key work is Julian Jaynes 1977. See also Campbell 1986. Friedman (1990) presents the fascinating results of experiments about how we perceive, locate, and construct time in a dimensional manner.

p. 156, **Zerubavel has traced** . . . : This paragraph and the following two are based on his 1985 book.

p. 161, **The earliest Romans** . . . : Sources of material in this paragraph are Aveni 1989, pp. 109–115, and Whitrow 1988, pp. 66–68.

p. 164, **. . . weekends terminate the week** . . . : Material in this and the previous paragraph come primarily from Zerubavel (1985), who explores the sociology of the week that pulses between workweek and weekend. I've also worked in a phrase from the title of Rybcyznski 1991.

p. 165, **Near year's end the Japanese** . . . : Rosemary Buffington, pers. comm.

p. 166, **biggest arrow of our calendars** . . . : An excellent discussion of the counts of various cultures is in Fraser 1987, pp. 91–95. This book is must-read for anyone interested in time. Data in this section for starts of counts is also from Whitrow 1988.

p. 169, **reset the time with a date acceptable to all:** Emiliani (1993) proposed the end of the last ice age (start of the Holocene) as an event that could satisfy. I (1994) pointed out problems and suggested my proposal herein.

p. 169, **signal was beamed from the Eiffel Tower** . . . : Kern 1983.

p. 170, **issues of how, where, and why to set the patterns of calendars** . . . : For in-depth critiques of the maps of time to which we conform our societies and ourselves, see Young 1988, Fraser 1987, and Rifkin 1987. Starhawk (1993) boldly imagined a multi-worshiping society in the not-so-distant future with rituals and festivals, based on creatively reworked traditions and new nature-based themes.

8. ARROWS

p. 175, **Ayla and Jondalar** . . . : Grand Paleolithic fiction by Jean Auel (1991)

p. 177, **. . . the page where he crossed out this guess:** Drake 1975.

p. 178, **thermal time . . . photothermal time** . . . : See Ritchie 1991.

p. 178, **systems theorist Donella Meadows** . . . : In Meadows et al. 1992, p. 17.

p. 179, **Laurie runs** . . . : Concepts in the next few paragraphs were inspired by what Lakoff and Johnson (1980) called the path metaphor.

p. 180, **Robert Frost:** His poem "The Road Not Taken," 1915.

p. 180, **Emily Dickinson:** Her poem "Because I Could Not Stop for Death," 1863.

p. 180, **"composing a life":** See M. C. Bateson 1989.

p. 181, **flower bud of a morning glory . . .** : Raab and Koning 1988.

p. 184, **"Port: Faith, sir, we were carousing . . . "**: *Macbeth* (Act 2, Scene 3).

p. 185, **paradox of New York City . . .** : Ruthen 1993.

p. 189, **"The rest of the phenomena . . . "**: Newton (from preface of first edition of *Principia Mathematica*), quoted in Adler 1955, ch. 10.

p. 189, **Goethe also envisioned . . .** : Gould (1991) noted Goethe's thesis for plant growth. I translated the couplet about God, but have been unable to relocate the original source.

p. 190, **physicists have listed . . .** : Hawking discussed three big arrows (1988, ch. 9). Cramer (1988) examined five. Penrose had seven (Rothman 1987). Morris (1985) had five. A philosopher (Horwich 1988) considered a constellation of ten asymmetries in philosophy and psychology. For an overview of thermodynamics and creative evolution, see Coveney and Highfield 1990. Swenson and Turvey (1991) argue for entropy being *the* arrow pulling life toward complexification. Big arrows are also integral to Gell-Mann's ideas (1994).

p. 191, **Biologists . . . engaged in debates . . .** : Cogent pieces by the biggest thinkers were assembled by Barlow (1994).

p. 191, **image of a bush . . .** : Gould 1989.

p. 192, **three large-scale . . . trends . . .** : Bonner 1988.

p. 192, **. . . negentropy:** Schrödinger 1945.

p. 192, **. . . syntropy . . .** : Fuller 1992.

p. 192, **"superstructuration":** Soleri 1969.

p. 192, **"the river that flows uphill":** Calvin 1986.

p. 192, **Japanese national anthem . . .** : Reed and Bristow 1993.

p. 193, **"Darwinian two-step":** Phrase from Calvin 1990, p. 280.

p. 193, **Because life had to begin simple . . .** : Concept from Gould 1988.

p. 194, **parts combine into wholes:** Maturana and Varela (1988) describe cells, organisms, and societies as first-, second-, and third-order unities in the evolution of complexity. Stock (1993) sees his Metaman as the fourth great combinatorial jump in complexity, after the bacterial cell, the eukaryotic cell, and the multicelled organism. Gell-Mann (1994, pp. 241–242) calls the formation of new levels via composites "the most fascinating increases in complexity."

p. 194, **. . . prokaryotes . . . symbioses . . .** : Margulis 1993. She uses the term symbiogenesis for parts combining functionally into wholes in biology.

p. 195, **merger of humans into societies . . .** : The functional differentiation of society is key to the Metaman concept (Stock 1993).

p. 195, **"It may be that some future observer . . . "**: Halle 1977, p. 437.

9. BREAKS

p. 198, **Thomas Kuhn's model** . . . : Kuhn 1962.

p. 198, **T. S. Eliot mused** . . . : His poem "The Hollow Men," 1925.

p. 199, **The clouds part** . . . : Wordsworth's poem "A Night Piece," 1798.

p. 200, **Chaos dynamicists** . . . : For an overview, see Gleick 1987.

p. 202, **breaks in time** . . . **similar to borders in space:** A close relation is evident throughout Zerubavel's (1993) superb examination of the sociology and psychology of distinctions.

p. 203, **pygmy chimps, orangutans,** . . . : Diamond 1992.

p. 203, **!Kung circle dance** . . . : Konner (1990) had a gripping firsthand experience (chapter called "Transcendental Medication").

p. 206, . . . **Badger dug it larger** . . . : Carr 1979.

p. 206, **ceremony of noren-wake** . . . : Kubiak 1990.

p. 206, **"plunging into the sea":** *New York Times*, April 6, 1993, p. A6.

p. 206, **I molded the clay into a pot** . . . : Once again, I am indebted to the insights of Lakoff and Johnson (1980).

p. 209, **"berserk":** See Eliade 1975.

p. 209, **"Clearly, clearly the dye** . . . ":** Borgatti 1983.

p. 210, **gates of dreaming:** Castenada 1993.

p. 213, **"suns" of the Aztecs** . . . : See Aveni 1989.

p. 213, **English language. . . broken by upheavals** . . . : Bryson 1990.

p. 213, **English Renaissance** . . . : A survey of attempts to number life's stages is in Chew 1962, ch. 6.

p. 217, **Einstein:** Story from Pais 1982, p. 178.

p. 217, **Schmandt-Besserat:** Story from Morrison 1992.

p. 217, **Lovelock:** Story from Lewin 1992, p. 115.

p. 218, . . . **Europe's "great thaw":** Clark 1969, ch. 2.

p. 218, **"rhythms of awakening":** Houston 1982.

p. 218, **"This is the Age wherein (methinks)** . . . ":** In Henry Power 1664, which I found when perusing the Folger Shakespeare Library in Washington, D.C.

p. 220, **"we force the spring":** From the 1993 Inaugural Address of President Clinton, *New York Times*, January 21, 1993.

p. 220, **Delineating the current new age** . . . : Some examples are drawn from the nearly one hundred cited societal transformations since 1950, compiled in Beniger 1986, pp. 4–5.

p. 220, . . . **Age of Bacteria:** Gould 1993.

p. 222, **"Spent and unthinking** . . . ":** Houston 1982, p. 186. Her quotations in the following paragraph are from Houston 1982, p. xvii, and 1990, p. 24.

p. 223, **The Millennial Project:** Savage 1992.

p. 223, **The Wildlands Project:** Updates about this exciting work appear regularly in *Wild Earth* magazine.

p. 223, **. . . partnership model of society . . . :** Eisler 1987 and Eisler and Combs 1992.

p. 223, **. . . death control . . . :** J. T. Fraser's (1987) vision of the world, with its "time-compact globe," the "global present," the "perceived irrelevance of history," and the "greying of the calendar," is well worth seeking out. He has developed a version of stratified stability in time with five levels.

p. 223, **"There's almost no myth I can think of . . . ":** Houston 1990, p. 35.

p. 223, **the Necker cube . . . :** The exercise showing its unstoppable transformations and the quotations about consciousness are from Pöppel 1988, ch. 7. Crick (1994) argues that solving the switching behavior of the Necker cube could be a key to consciousness, because the switch must occur at some stage subsequent to retinal sensing, thus closing in on the neurons that are interpretive (i.e., close to "mind").

10. CYCLES

p. 228, **our species' internal biological clock . . . :** Six cave-ensconced subjects had an average of 24 hours and 42 minutes (Young 1988, p. 33). This length, however, can vary among individuals by as much as several hours (Fraser 1987, p. 119).

p. 228, **. . . to the tenth of a day:** Fraser 1987, p. 120.

p. 229, **swan's legs . . . :** French 1988, pp. 198–199.

p. 230, **"genius of recycling . . . ":** Allen and Hoekstra 1992, p. 253.

p. 230, **circles of influence, circles of causation . . . :** Fuzzy logic explores political, economic, and social issues using network diagrams called fuzzy cognitive models (Kosko 1993).

p. 237, **. . . carbonates are heat-shattered . . . :** Subduction of significant amounts of carbonates probably required the evolution of the planktonic carbonate-shelled organisms, such as coccolithophorids and foraminifera. This evolutionary event during the last 200 million years would have perturbed the carbon cycle by increasing the degassing of carbon dioxide from the subduction zones (Volk 1989).

p. 238, **Howard Odum . . . :** Odum 1983.

p. 238, **James Beniger . . . :** Beniger 1986, pp. 215–216.

p. 238, **"great stochastic process":** See Bateson 1979, ch. 6.

p. 238, **Richard Dawkins contends . . . :** Dawkins 1982.

p. 239, **"The Method (therefore) . . . ":** Hooke quoted in Derham 1726, pp. 26–28.

p. 241, **Is biological evolution an arrow?:** Follow the cross fire with the most prominent debaters of this issue during the last hundred years in Barlow 1994.

p. 241, **A Hindu tale . . . :** Zimmer 1946.

p. 241, **Ozymandius:** Poem by Shelley, 1817.

p. 242, **Starlings are sophisticated songsters:** Adret-Hausberger and Jenkins 1988.

p. 242, **time patterns in a Beethoven symphony . . . :** Evans 1923.

p. 245, **Life is an arrow filled with . . . habits:** Timesmith Michael Young (1988) calls habit the "flywheel of society."

p. 245, **"When evening approached . . . ":** This and other Rousseau quotations come from Rousseau 1979, pp. 62–73. The "Fifth Walk" was probably written in 1777.

References

Ackerman, Diane. 1990. *A Natural History of the Senses.* New York: Random House.

Addiss, Stephen. 1989. *The Art of Zen.* New York: Harry N. Abrams.

Adler, Mortimer J., ed. 1955. *The Great Ideas: A Syntopicon of Great Books of the Western World.* Chicago: Encyclopedia Brittanica.

Adret-Hausberger, M. and Peter F. Jenkins. 1988. Complex organization of the warbling song in the European starling *Sturnus vulgaris. Behaviour* 107:138–156.

Allen, Timothy F. H. and Thomas W. Hoekstra. 1992. *Toward a Unified Ecology.* New York: Columbia University Press.

————. 1984. Nested and non-nested hierarchies: A significant distinction for ecological systems. In A. W. Smith, ed., *Systems Methodologies and Isomorphies*, Proceedings of the 24th annual meeting of S.G.S.R. Seaside, Calif.: Intersystem.

Allen, Timothy F. H. and Thomas B. Starr. 1982. *Hierarchy: Perspectives for Ecological Complexity.* Chicago: University of Chicago Press.

Auel, Jean. 1991. *The Plains of Passage.* New York: Bantam.

Aveni, Anthony F. 1989. *Empires of Time: Calendars, Clocks, and Cultures.* New York: Basic Books.

Bancroft, Anne. 1991. *Zen: Direct Pointing to Reality.* New York: Thames and Hudson.

Barlow, Connie. 1994. *Evolution Extended: Biological Debates on the Meaning of Life.* Cambridge: MIT Press.

————. 1991. *From Gaia to Selfish Genes: Selected Writings in the Life Sciences.* Cambridge: MIT Press.

Barlow, Connie and Tyler Volk. 1992. Gaia and evolutionary biology. *BioScience* 42:686–693.

———. 1990. Open systems living in a closed biosphere: A new paradox for the Gaia debate. *BioSystems* 23:371–384.

Bateson, Gregory. 1979. *Mind and Nature: A Necessary Unity.* New York: Dutton.

Bateson, Gregory and Mary Catherine Bateson. 1988. *Angels Fear: Toward an Epistomology of the Sacred.* New York: Bantam.

Bateson, Mary Catherine. 1989. *Composing a Life.* New York: Atlantic Monthly Press.

Benedict, Ruth. 1992. The synergy lectures. In A. Combs, ed., *Cooperation: Beyond the Age of Competition*, pp. 58–67. Philadelphia: Gordon and Breach.

Beniger, James R. 1986. *The Control Revolution: Technological and Economic Origins of the Information Society.* Cambridge: Harvard University Press.

Blerkom, Jonathon van and George Henry. 1991. Dispermic fertilization of human oocytes. *Journal of Electron Microscopy Technique* 17:437–449.

Bly, Robert. 1990. *Iron John: A Book About Men.* Reading, Mass.: Addison-Wesley.

Bonner, John Tyler. 1988. *The Evolution of Complexity by Means of Natural Selection.* Princeton: Princeton University Press.

Boorstin, Daniel J. 1985. *The Discoverers.* New York: Vintage.

Borgatti, Jean. 1983. *Cloth as a Metaphor: Nigerian Textiles from the Museum of Cultural History.* Los Angeles: University of California Press (The Museum).

Bronowski, Jacob. 1974. *The Ascent of Man.* Boston: Little, Brown.

———. 1970. New concepts in the evolution of complexity: Stratified stability and unbounded plans. *Zygon* 5:18–35.

Brown, Joseph Epes. 1989/1953. *The Sacred Pipe: Black Elk's Account of the Seven Rites of the Oglala Sioux.* Norman, Okla.: University of Oklahoma Press.

Bryson, Bill. 1990. *The Mother Tongue: English and How It Got that Way.* New York: William Morrow.

Calvin, William H. 1990. *The Cerebral Symphony: Seashore Reflections on the Structure of Consciousness.* New York: Bantam.

———. 1986. *The River that Flows Uphill: A Journey from the Big Bang to the Big Brain.* New York: Macmillan.

Campbell, Jeremy. 1986. *Winston Churchill's Afternoon Nap: A Wide-Awake Inquiry into the Human Nature of Time.* New York: Simon and Schuster.

Campbell, Joseph. 1974. *The Mythic Image.* Princeton: Princeton University Press.

———. 1973/1949. *The Hero with a Thousand Faces.* Princeton: Princeton University Press.

Carr, Pat. 1979. *Mimbres Mythology.* El Paso: Texas Western Press.

Castenada, Carlos. 1993. *The Art of Dreaming.* New York: HarperCollins.

Casti, John L. 1994. *Complexification: Explaining a Paradoxical World Through the Science of Surprise.* New York: HarperCollins.

Chew, Samuel C. 1962. *The Pilgrimage of Life.* New Haven: Yale University Press.

Clark, Kenneth. 1969. *Civilization.* New York: Harper and Row.

Cohen, Jack and Ian Stewart. 1994. *The Collapse of Chaos: Discovering Simplicity in a Complex World.* New York: Viking.

Copernicus, Nicholas (Nicolaus). 1992/1543. *On the Revolutions.* Translation and commentary by Edward Rosen. Baltimore: John Hopkins University Press.

X Coveney, Peter and Roger Highfield. 1990. *The Arrow of Time.* New York: Fawcett Columbine.

Cramer, John G. 1988. Velocity reversal and the arrows of time. *Foundations of Physics* 18:1205–1212.

Crick, Francis. 1994. *The Astonishing Hypothesis: The Scientific Search for the Soul.* New York: Charles Scribner's Sons.

Dawkins, Richard. 1987. *The Blind Watchmaker.* New York: Norton.

———. 1982. *The Extended Phenotype.* New York: Freeman.

Dennett, Daniel C. 1991. *Consciousness Explained.* Boston: Little, Brown.

Derham, W. 1726. *Philosophical Experiments and Observations of the Late Eminent Dr. Robert Hooke.* London: W. and J. Innys.

Diamond, Jared. 1992. *The Third Chimpanzee: The Evolution and Future of the Human Animal.* New York: HarperCollins.

Drake, Stillman. 1975. The role of music in Galileo's experiments. *Scientific American,* June, pp. 98–104.

Dumoulin, Heinrich. 1963. *A History of Zen Buddhism.* New York: Pantheon.

Durant, Will. 1943/1926. *The Story of Philosophy.* Garden City, N.Y.: Garden City Publishing.

Eisler, Riane. 1987. *The Chalice and the Blade.* New York: HarperCollins.

Eisler, Riane and Allan Combs. 1992. Cooperation, competition and gylany: Cultural evolution from a new dynamic perspective. In A. Combs, ed., *Cooperation: Beyond the Age of Competition,* pp. 75–85. Philadelphia: Gordon and Breach.

Eliade, Mircea. 1975. *Rites and Symbols of Initiation.* New York: Harper and Row.

———. 1971. *The Forge and the Crucible: The Origins and Structures of Alchemy.* New York: Harper and Row.

X Emiliani, C. 1993. Calendar reform. *Nature* 366:716.

Evans, Edwin. 1923. *Beethoven's Nine Symphonies Fully Described and Analyzed.* London: W. R. Reeves.

Flugel, J. C. 1950. *The Psychology of Clothes.* London: Hogarth Press.

Fraser, J. T. 1987. *Time the Familiar Stranger.* Amherst: University of Massachusetts Press.

———. 1975. *Of Time, Passion, and Knowledge: Reflections on the Strategy of Existence.* New York: George Braziller.

Freeman, Norman H. 1987. Children's drawings of human figures. In R. L. Gregory, ed., *The Oxford Companion to the Mind*, pp. 135–139. Oxford: Oxford University Press.

French, X. 1988. *Evolution and Invention: Design in Nature and Engineering.* Cambridge, U.K.: Cambridge University Press.

Friedman, William J. 1990. *About Time: Inventing the Fourth Dimension.* Cambridge: MIT Press.

Fryer, Jonathan. 1975. *The Great Wall of China.* London: New English Library.

Fuller, R. Buckminster, with Kiyoshi Kuromiya, adjuvant. 1992. *Cosmography: A Posthumous Scenario for the Future of Humanity.* New York: Macmillan.

———. 1981. *Critical Path.* New York: St. Martin's Press.

Gell-Mann, Murray. 1994. *The Quark and the Jaguar: Adventures in the Simple and the Complex.* New York: Freeman.

Giedion, S. 1971. *Architecture and the Phenomena of Transition: The Three Space Conceptions in Architecture.* Cambridge: Harvard University Press.

Gilligan, Carol. 1982. *In a Different Voice: Psychological Theory and Women's Development.* Cambridge: Harvard University Press.

Gleick, James. 1987. *Chaos.* New York: Viking Penguin.

Gore, Al. 1992. *Earth in the Balance: Ecology and the Human Spirit.* Boston: Houghton Mifflin.

Gould, Stephen Jay. 1993. Prophet for the earth [review of Wilson 1992]. *Nature* 361:311–312.

———. 1991. More light on leaves. *Natural History* 2/91:16–23.

———. 1989. *Wonderful Life: The Burgess Shale and the Nature of History.* New York: Norton.

———. 1988. Trends as changes in variance: A new slant on progress and directionality in evolution. *Journal of Paleontology* 62:319–329.

———. 1987. *Time's Arrow, Time's Cycle: Myth and Metaphor in the Discovery of Geological Time.* Cambridge: Harvard University Press.

Greulach, Victor A. 1973. *Plant Function and Structure.* New York: Macmillan.

Halevi, Z'ev ben Shimon. 1988/1979. *Kabbalah: Tradition of Hidden Knowledge.* New York: Thames and Hudson.

Halifax, Joan. 1982. *Shaman: The Wounded Healer.* New York: Thames and Hudson.

Halle, Louis J. 1977. *Out of Chaos.* Boston: Houghton Mifflin.

Hardison, O. B., Jr. 1989. *Disappearing Through the Skylight: Culture and Technology in the Twentieth Century.* New York: Viking.

Harjo, Joy. 1989. *Secrets from the Center of the World*. Tucson: University of Arizona Press.

Harner, Michael. 1990/1980. *The Way of the Shaman*. New York: HarperCollins.

Hawking, Stephen W. 1988. *A Brief History of Time: From the Big Bang to Black Holes*. New York: Bantam.

Hern, Warren. M. 1993. Is human culture carcinogenic for uncontrolled population growth and ecological destruction? *BioScience* 43:768–773.

Hook, B., ed. 1982. *Cambridge Encyclopedia of China*. Cambridge, U.K.: Cambridge University Press.

Horney, Karen. 1945. *Our Inner Conflicts*. New York: W. W. Norton.

Horwich, Paul. 1988. *Asymmetries in Time: Problems in the Philosophy of Science*. Cambridge: MIT Press.

Houston, Jean. 1992. *The Hero and the Goddess: The Odyssey as Mystery and Initiation*. New York: Ballantine.

———. 1990. The promise of the new millenium (a dialogue). *Anima* 17/1:19–44.

———. 1982. *The Possible Human: A Course in Extending Your Physical, Mental, and Creative Abilities*. Los Angeles: J. P. Tarcher.

Hutchins, Robert Maynard, ed. 1952. *Great Books of the Western World*. Vol. 7—The Dialogues of Plato. Vol. 16—Ptolemy, Copernicus, Kepler. Chicago: Encyclopaedia Britannica.

Huxley, Aldous. 1977. *Moksha: Writings on Psychedelics and the Visionary Experience (1931–1963)*. M. Horowitz and C. Palmer, eds. New York: Stonehill.

———. 1968. *Island*. New York: Bantam.

———. 1954. *The Doors of Perception*. New York: Harper and Row.

Jaynes, Julian. 1977. *The Origin of Consciousness in the Breakdown of the Bicameral Mind*. Boston: Houghton Mifflin.

Johnson, Buffie. 1988. *Lady of the Beasts: Ancient Images of the Goddess and Her Sacred Animals*. New York: Harper and Row.

Johnson, Mark. 1987. *The Body in the Mind: The Bodily Basis of Meaning, Imagination, and Reason*. Chicago: University of Chicago Press.

Jung, Carl. G. 1978. *Psychology and Alchemy*. Princeton: Princeton University Press.

Kelly, Kevin. 1994. *Out of Control: The Rise of Neo-Biological Civilization*. Reading, Mass.: Addison-Wesley.

Kern, Stephen. 1983. *The Culture of Time and Space 1880–1918*. Cambridge: Harvard University Press.

Keswick, Maggie. 1986. *The Chinese Garden: History, Art, and Architecture*. New York: St. Martin's Press.

Koestler, Arthur. 1979. *Janus: A Summing Up*. New York: Vintage.

———. 1967. *The Ghost in the Machine*. New York: Macmillan.

Konner, Melvin. 1990. *Why the Reckless Survive and Other Secrets of Human Nature*. New York: Viking.

Kosko, Bart. 1993. *Fuzzy Thinking: The New Science of Fuzzy Logic*. New York: Hyperion.

Kubiak, W. David. 1990. E pluribus Yamato: The culture of corporate beings. *Whole Earth Review*, Winter, pp. 4–10.

Kuhn, Thomas S. 1962. *The Structure of Scientific Revolutions*. Chicago: University of Chicago Press.

LaBarbera, Michael. 1990. Principles of design in fluid transport systems in zoology. *Science* 249:992–1000.

Lakoff, George and Mark Johnson. 1980. *Metaphors We Live By*. Chicago: University of Chicago Press.

Lehmann, Karl. 1945. The dome of heaven. *The Art Bulletin*, March, 1–27.

Lethaby, William R. 1975. *Architecture, Mysticism, and Myth*. New York: George Braziller.

Lewin, Roger. 1992. *Complexity: Life at the Edge of Chaos*. New York: Macmillan.

Logan, Robert K. 1986. *The Alphabet Effect: The Impact of the Phonetic Alphabet on the Development of Western Civilization*. New York: William Morrow.

Lovelock, James E. 1988. *The Ages of Gaia: A Biography of Our Living Earth*. New York: Norton.

———. 1979. *Gaia: A New Look at Life on Earth*. Oxford: Oxford University Press.

Mach, Edmund von, ed. 1916. *Greek and Roman Sculpture* (series A). Boston: The University Prints.

Mankiewicz, Paul S. 1991. The macromolecular matrix of plant cell walls as a major Gaian interfacial regulator in terrestrial environments. In S. H. Schneider and P. J. Boston, eds., *Scientists on Gaia*, pp. 309–319. Cambridge: MIT Press.

Margulis, Lynn. 1993/1981. *Symbiosis in Cell Evolution*. New York: Freeman.

———. 1992. Power to the protoctists. *Earthwatch* Sept./Oct.: 25–29.

———. 1988. *Five Kingdoms*. New York: Freeman.

Maturana, Humberto R. and Francisco J. Varela. 1988. *The Tree of Knowledge: The Biological Roots of Human Understanding*. Boston: Shambhala.

McCarty, Maclyn. 1986. *The Transforming Principle: Discovering that Genes are Made of DNA*. New York: Norton.

McLean, Adam. 1989. *The Triple Goddess: An Exploration of the Archetypal Feminine*. Grand Rapids, Mich.: Phanes.

Meadows, Donella H., Dennis L. Meadows, and J. Randers. 1992. *Beyond the Limits: Confronting Global Collapse and Envisioning a Sustainable Future (Executive Summary)*. Post Mills, Vermont: Chelsea Green.

Miller, G. A. 1956. The magical number seven, plus or minus two. *Psychological Review* 63:81–97.

Miller, James G. 1978. *Living Systems.* New York: McGraw-Hill.

Ming, D. W. and D. L. Henninger, eds. 1989. *Lunar Base Agriculture: Soils for Plant Growth*. Madison, Wisc.: American Society of Agronomy.

Moore, John C. and H. William Hunt. 1988. Resource compartmentation and the stability of real ecosystems. *Nature* 333:261–263.

Moore, Peter D. 1992. Not only gone with the wind. *Nature* 357:116.

Morgan, W. N. 1980. *Prehistoric Architecture in the Eastern United States*. Cambridge: MIT Press.

Morris, Richard. 1985. *Time's Arrows: Scientific Attitudes Toward Time*. New York: Simon and Schuster.

Morrison, Philip. 1992. Three-dimensional words, review of *Before Writing*, by Denise Schmadt-Besserat. *Scientific American*, November, pp. 133–134.

Odum, Howard T. 1983. *Systems Ecology: An Introduction*. New York: John Wiley and Sons.

Pais, Abraham. 1982. *Subtle is the Lord: The Science and the Life of Albert Einstein*. Oxford: Oxford University Press.

Payne, Blance. 1965. *History of Costume: From the Ancient Egyptians to the Twentieth Century*. New York: Harper and Row.

Pirsig, Robert M. 1985/1974. *Zen and the Art of Motorcycle Maintenance: An Inquiry into Values*. New York: Bantam.

Pöppel, Ernst. 1988. *Mindworks: Time and Conscious Experience*. Boston: Harcourt Brace Jovanovich.

Power, Henry. 1664. *Experimental Philosophy, in Three Books: Containing New Experiments Microscopical, Mercurial, Magnetical*. London: T. Roycroft.

Prete, Frederick R. 1991. Can females rule the hive? The controversy over honey bee gender roles in British beekeeping texts of the sixteenth–eighteenth centries. *Journal of the History of Biology* 24:113–144.

Raab, Mandy M. and Ross E. Koning. 1988. How is Floral Expansion Regulated? *BioScience* 38:670–674.

Reed, W. L. and M. J. Bristow. 1993. *National Anthems of the World*. London: Cassell.

Reeves, Hubert. 1991. *The Hour of Our Delight: Cosmic Evolution, Order, and Complexity*. New York: Freeman.

Rifkin, Jeremy. 1987. *Time Wars: The Primary Conflict in Human History*. New York: Henry Holt.

Ritchie, J. T. 1991. Wheat phasic development. In J. Hanks and J. T. Ritchie, eds., *Modeling Plant and Soil Systems—Agronomy Monograph no. 31*, pp. 31–54. Madison, Wisc.: American Society of Agronomy.

⅄ Rothman, Tony. 1987. Seven Arrows of Time. *Discover*, February, pp. 63–77.

Rosenau, H. 1976. *Boullée and Visionary Architecture*. London: Academy Editions.

⅄ Rousseau, Jean-Jacques. 1979. *The Reveries of the Solitary Walker*. New York: New York University Press.

Rule, John C. 1974. *Louis XIV*. Englewood Cliffs, N.J.: Prentice-Hall.

Ruthen, Russel. 1993. Adapting to complexity. *Scientific American*, January, pp. 128–140.

Rybczynski, Witold. 1991. *Waiting for the Weekend*. New York: Viking.

Savage, Marshall T. 1992. *The Millennial Project: Colonizing the Galaxy—In Eight Easy Steps*. Denver: Empyrean.

Schele, Linda and David Freidel. 1990. *A Forest of Kings: The Untold Story of the Ancient Maya*. New York: William Morrow.

Schneider, W. 1963. *Babylon is Everywhere: The City as Man's Fate*. London: McGraw-Hill.

Schopf, J. William, ed. 1983. *Earth's Earliest Biosphere: Its Origin and Evolution*. Princeton: Princeton University Press.

Schrödinger, Erwin. 1945. *What Is Life?* New York: Macmillan.

Schwartzman, David W. and Tyler Volk. 1989. Biotic enhancement of weathering and the habitability of Earth. *Nature* 340:457–460.

Simon, Herbert A. 1981/1969. *The Sciences of the Artificial*. Cambridge: MIT Press.

Smith, E. Baldwin. 1971/1950. *The Dome: A Study in the History of Ideas*. Princeton: Princeton University Press.

Soellner, Rolf. 1984. *King Lear* and the magic of the wheel. *Shakespeare Quarterly* 35:274–289.

Soleri, Paolo. 1969. *Arcology: The City in the Image of Man*. Cambridge: MIT Press.

Starhawk. 1993. *The Fifth Sacred Thing*. New York: Bantam.

Stock, Gregory. 1993. *Metaman: The Merging of Humans and Machines into a Global Superorganism*. New York: Simon and Schuster.

Sutton, Peter, ed. 1988. *Dreamings: The Art of Aboriginal Australia*. New York: George Braziller.

Swenson, Rod and M. T. Turvey. 1991. Thermodynamic reasons for perception-action cycles. *Ecological Psychology* 3:317–348.

Tannen, Deborah. 1990. *You Just Don't Understand: Women and Men in Conversation*. New York: William Morrow.

Taubes, Gary. 1991. The disputed birth of buckyballs. *Science* 253:1476–1479.

Thompson, D'Arcy Wentworth. 1971/1917. *On Growth and Form.* Cambridge, U.K.: Cambridge University Press.

Vogel, Steven. 1988. *Life's Devices: The Physical World of Animals and Plants.* Princeton: Princeton University Press.

Volk, Tyler. 1994. Re-setting the great count. *Nature* 370:244.

———. 1989. Sensitivity of climate and atmospheric CO_2 to deep-ocean and shallow-ocean carbonate burial. *Nature* 337:637–640.

———. 1987a. Rise of angiosperms as a factor in long-term climatic cooling. *Geology* 17:107–110.

———. 1987b. Feedbacks between weathering and atmospheric CO_2 over the last 100 million years. *American Journal of Science* 287:763–779.

Wainwright, Stephen A. 1988. *Axis and Circumference: The Cylindrical Shape of Plants and Animals.* Cambridge: Harvard University Press.

Waldrop, M. Mitchell. 1992. *Complexity: The Emerging Science at the Edge of Order and Chaos.* New York: Simon and Schuster.

Watson, James D. 1968. *The Double Helix: A Personal Account of the Discovery of DNA.* New York: Atheneum.

Whitrow, G. J. 1988. *Time in History: The Evolution of Our General Awareness of Time and Temporal Perspective.* Oxford: Oxford University Press.

Wilbur, Ken. 1981. *No Boundary: Eastern and Western Approaches to Personal Growth.* Boston: New Science Library.

Williamson, Arthur H. 1988. The cultural meanings of science. *Interchange* 2, issue 2.

Wilson, Edward O. 1992. *The Diversity of Life.* Cambridge: Harvard University Press.

Wind, Edgar. 1967. *Pagan Mysteries in the Renaissance.* Middlesex, U.K.: Penguin.

Yates, Frances A. 1966. *The Art of Memory.* Chicago: University of Chicago Press.

Young, Michael. 1988. *The Metronomic Society: Natural Rhythms and Human Timetables.* Cambridge: Harvard University Press.

Zerubavel, Eviatar. 1993/1991. *The Fine Line: Making Distinctions in Everyday Life.* Chicago: University of Chicago Press.

———. 1985. *The Seven Day Circle: The History and Meaning of the Week.* New York: Free Press.

Zimmer, Heinrich. 1946. *Myths and Symbols in Indian Art and Civilization.* New York: Pantheon.

Illustration Credits

Figures or parts of figures not acknowledged here are given credit in the captions or were supplied by the author. The latter include photographs taken during travels, hand and computer drawings, and organisms digitized on a flatbed scanner (no animals were killed or harmed when this art was made). Sources no longer in copyright are cited if they may be of scientific or cultural interest and if the parts used are important in the composites; otherwise, such sources are designated OB (for "old book"). A very few instances for which sources have been lost are designated SL. Omissions brought to our attention for these or others will be corrected in future editions.

Sources that are cited more than once are abbreviated as follows:

(LH)   courtesy of Lyn Hughes
(CB)   courtesy of Connie Barlow
(JWP1) Powell, J. W. 1893. *Tenth Annual Report of the United States Bureau of Ethnology, 1888–1889*. Washington: United States Government.
(JWP2) Powell, J. W. 1894. *Eleventh Annual Report of the United States Bureau of Ethnology, 1889–1890*. Washington: United States Government.
(DTS) Suzuki, Daisetz Teitaro. 1960. *Manual of Zen Buddhism*. New York: Grove/Atlantic Press. Courtesy of Grove/Atlantic.

1. SPHERES

In the beginning is the sphere. Ovum, fetus (OB).
Spheres at all levels. Atoms of germanium (courtesy of John J. Boland. See his 1992 article, "Hydrogen as a probe of semiconductor surface structure:

The Ge(111)-c(2x8) surface." *Science* 255: 186–188). Radiolarian, pollen (OB). Earth (NASA).

Play ball. Native American ballplayer on incised shell (P. Phillips and J. A. Brown, *Pre-Columbian Shell Engravings from the Craig Mound at Spiro, Oklahoma*, vol. 1, Peabody Museum Press. Copyright 1975 by the President and Fellows of Harvard College).

Rolling, tumbling, shooting spheres: Dung beetles (OB). *Vanguard 1* satellite (courtesy of the National Air and Space Museum, Smithsonian Institution).

Sphericity in space: NASA photographs 51F-12-0032 and 51F-16-0013.

A pile of domes: Tortoise (OB). Buckyball (structure is by Steven R. Wilson; image by Robert Lancaster; reproduced courtesy of *Academic Computing and Networking*, New York: New York University).

Inside the beehive tomb of Atreus: Author in empty tomb of Atreus (LH). Pregnant woman in front of United States Capitol (courtesy of Ken and Michelle Volk).

Spheres in the hands and heads of the mythic imagination: Australian pictograph head (JWP1, p. 489). Christian head, Cleopatra (OB). Buddhist head (DTS, p. 160). Buddhist goddess (courtesy of John Powers, from p. 98 of John M. Rosenfield and Shujiro Shimada, 1970, *Traditions of Japanese Art, Selections from the Kimiko and John Powers Collection*, Cambridge, Mass.: Fogg Art Museum).

Spheres in the hands of science: Muse of science (Corbould). Rainbow and droplets (S'Gravesande, Wilhelm Jacob. 1722. *Physices elementa mathematica experimentis confirmata*). Candle (Huygens, Christian. 1728. *Opera reliquia*). Vacuum sphere (Guericke, Otto von. 1672. *Experimenta nova Mageburgica de vacuo spatio*). Lodestone (Gilbert, William. 1600. *De magnete*).

2. TUBES AND SHEETS

Physical, biological, and cultural sheets: Floor of Gran Trianon (LH).

Sheets for transferring matter, energy, and messages: Trout lamellae (courtesy of National Research Council of Canada, from P. Laurent and N. Hebibi, 1989, Gill morphometry and fish osmoregulation, *Canadian Journal of Zoology* 67: 3055–3063). Mayan stela (OB).

Tubes as transfer surfaces: Pine (CB). Tetrahymena (courtesy of Birgit Satir).

Structural tubes: Eiffel Tower (LH).

Tubes for transport: Nerves (OB).

Spheres and tubes as mental models: Australian bark painting ("Totemic Site" by Kianoo Tjeemairee, courtesy of the Aboriginal Artists Agency; figure 88 from Sutton et al. 1988). Kabbalah (OB). Poster diagram (courtesy of the American Society of Agronomy).

Sphere and tube as archetypal pair in art and architecture: Queen Elizabeth (engraving by Crispin de Passe).

Tubes and spheres in mind and nature: Fungus, foraminiferan, vessel, eye (OB).

3. BORDERS

Archetypal containers: Sia vessel (JWP2, pl. XXV). Chilkat shirt (JWP1, p. 428). Reims cathedral (OB).

Openings for creativity: Bark carving (JWP1, p. 242).

Nampeyo: (courtesy of Tad Nichols, no. PIX-1465 from the Arizona State Museum, Tucson).

Rectangles . . . : Beijing's walls (OB). Tennis match (LH).

4. BINARIES

Binaries in biology: Egg and sperms (OB).

Binaries in Daoism, alchemy, and physics: Sun and moon (OB).

Binary as conflict: Sea lions (LH). Chess (OB).

Binary as union: Moths (CB).

The earth & sky archetype: Buddha (DTS, p. 158). Christ the Judge, from Michaelangelo's *Last Judgment* (OB).

Interplay between two and one: Binary lion (Matthèus Merian the Elder, 1618).

Beyond the binary with threes: The Three Graces (Botticelli's *Primavera*). Three Erinyes (illustration by Gustav Doré for Dante's *Divine Comedy*, Canto 9 of *Inferno*).

Experiment and control: Flasks (Pasteur, Louis. 1861. *Annales des science naturelles*).

5. CENTERS

Royalty and sun: Louis XIV print (Bourgeois, Émile. 1895. *The Century of Louis XIV*. p. 156. London: Sampson Low, Marston, and Company). Egyptian sun-god blessing royalty (d'Avennes, Prisse. 1863. *Histoire de l'art Ägyptien d'apres les monuments*).

The nucleated atom: (Chadwick, James. 1963. *The Collected Papers of Lord Rutherford of Nelson*. London: George Allen and Unwin).

The nucleated cell: Amoeba (SL).

Nerve system of a scorpion: Scorpion (OB).

Government as social center: African village (courtesy of the American Geographical Society, from fig. 110 of Richard Upjohn Light, 1944/1941, *Focus on Africa*, New York: A.G.S.).

Nature as center: Spider web (CB). Spider experiment (after Rüdiger Paul and Till Fincke, 1989, Book lung function in arachnids: II. Carbon dioxide release and its relation to respiratory surface, water loss and heart frequency, *Journal of Comparative Physiology B* 159: 419–432).

6. LAYERS

Dante and Beatrice reach the vision of the rose: (illustration by Gustav Doré for Dante's *Divine Comedy*, Canto 31 of *Paradiso*).
Hierarchies: NSF (United States Government, 1990). Mosaic in mosque, Ankara, Turkey (courtesy of Hatice Cullingford).
Holarchies. Earth (NASA). Ciliate (OB). Four examples of rock art (JWP1, pp. 86, 120, 194).
Clonons: Cabbages (LH).
Handfuls of holons: Chinese elements (Chou Tun-I, 11th century). Ceremonial figures (JWP1, p. 511). Banner, head, ants (OB).

7. CALENDARS

Portraits of time: Detail (Brueghel, 16th century). Dakota symbols (JWP1, p. 265).
Beginning and ending years and days: Data (from sources in the reference list: Aveni 1989, Whitrow 1988, Boorstin 1985, Fraser 1975).

8. ARROWS

Galileo's experiment. Galileo (detail of painting by G. Martelli, 1841). Graphs (by author).
Arrows in property space. Data transformed by the author from various sources.
Teaching the arrows of time. Apples (courtesy of Nancy and Kevin Hartel).

9. BREAKS

Holy transformations: Sia masks (JWP2, pl. XXXI). Sipapu climbers (courtesy of Texas Western Press, from pl. 4 of Carr 1979).
Sequences of stages in psychology, geology, and biology: Zen stages (DTS, pl. 7–9). Geological time scale (after diagram from Geological Society of America).
Forks in life: The "Y" of Pythagorus (Heyns, Zacharias. 1625. *Emblemata*).
Human epiphanies: Unveiling of Truth (frontispiece of *Encyclopédie*, 1751–1780). Greek ritual (Mach, Edmund von, ed. 1916. *Greek and Roman Sculpture*. Boston: The University Prints).

Transformation of the psyche: Lizard and fire (Maier, Michael. 1618. *Atalanta fugiens . . .*). Balinese girl (courtesy of Institute for Intercultural Studies; from photograph No. 98 by Gregory Bateson [1930] in Jane Belo, 1960, *Trance in Bali*, New York: Columbia University Press).

10. CYCLES

Sleep-wake patterns of an infant: Infant sleep pattern (courtesy of the American Physiological Society, from Nathaniel Kleitman and Theodore G. Engelmann, 1953, Sleep characteristics of infants, *Journal of Applied Physiology* 7: 269–282).

Hurricane dancers: spins in physical space: Hurricane (NASA). Dancers (after a painting by William Blake for *A Midsummer Night's Dream*).

Cycles of causation and control: Engineering cycle (courtesy of Joe Franceschi). Menstrual cycle (after page 57 of Sharon Golub, 1992, *Periods: From Menarche to Menopause*, Newbury Park, Calif.: Sage).

Two cycles in the property space of temperature: New York City temperature data transformed by author from data in *The New York Times*, 8 January 1989, 7 January 1990, 6 January 1991.

Binary cycles in arrows: Peruvian birch bark drawing of canoe (JWP1, pl. XVI).

Power cycles in the arrows of civilization: Windmill in Greece (LH).

Cycles, breaks, and arrows in music: Starling sonograph (courtesy of Peter F. Jenkins). Beethoven manuscript (courtesy of Columbia University Press).

Flower of Sacred Datura and human ovum: Ovum (OB).

Index